The Skinny on Soy

Marie Oser

The Skinny on Soy

We gratefully acknowledge the Soyfoods Association of North America (SANA) for granting the use of the decorative graphics in this book and for all that they do to promote the extensive health benefits of consuming soyfoods. For more information visit www.soyfoods.org.

Published by:
Vegan Publishers
Danvers, MA
www.veganpublishers.com

Cover photo: Brad Haskell; Food styling: Paty Winters

Page X photo: Brad Haskell; Food styling: Paty Winters

Page 127: Photo courtesy of the WISH Foundation, American Soybean Association

Page 129: Photo courtesy of the SoyInfo Center

About the Author photo: Brad Haskell

Cover and text design: Nicola May Design

Printed in the United States of America

ISBN: 9781940184357

CONTENTS

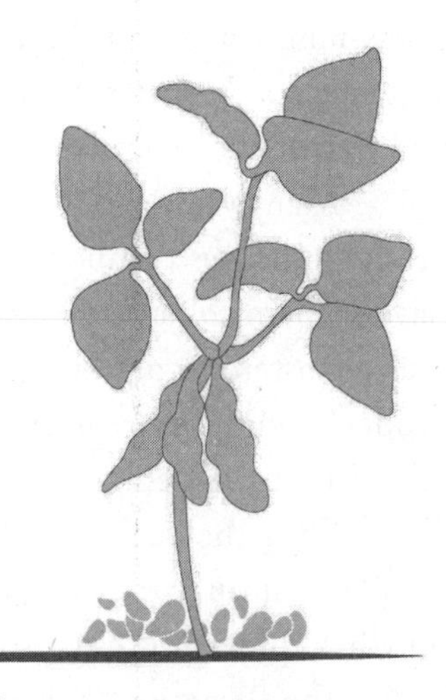

FOREWORD

BY MICHAEL GREGER, MD

Whether the myth has been kept alive by the dairy industry or simply by web trolls, the claim that consuming soy is detrimental to health has been remarkably persistent in recent years.

A dietary staple in numerous parts of the world, soybeans are among the most nutritious and easily digested of legumes.

There are so many reasons to love soy! Whole soyfoods such as tempeh and edamame, like all other legumes—beans, split peas, chickpeas, and lentils—are nutrient dense, offer affordable nutrition that can contribute to an optimal diet, and are included in the USDA dietary guidelines.

The federal government's MyPlate campaign was developed to prompt Americans to think about building healthy meals. Most of your plate should be covered with vegetables and grains (preferably whole grains) with the rest of the plate split between fruits and the protein group. Legumes like soy are given special treatment, straddling both the protein and the vegetable groups. They're loaded with protein, iron, and zinc, as you might expect from other protein sources like meat. However, legumes also contain nutrients that are concentrated in the vegetable kingdom, including fiber, folate, and potassium. You get the best of both worlds with beans, all the while enjoying foods that are naturally low in saturated fat and sodium and free of cholesterol.

Many years ago researchers found that soybeans contain naturally occurring compounds called isoflavones. These compounds have a chemical structure somewhat similar to human hormones, like testosterone or estrogen, leading some to speculate that soy products might exert hormonal effects, such as increasing the risk of cancer in women.

Over the years these biological issues have been studied in detail and it turns out that, instead of soy causing cancer, it may not only help prevent it but also prolong survival.

The allegation that consuming soyfoods causes cancer is only one of the many alarming claims addressed in *The Skinny on Soy.* Marie Oser has conducted a thorough review of the many issues raised by critics of this versatile and nutritious legume used in Asia for more than five thousand years.

It may be no coincidence that two populations with the longest lifespans, the Okinawa Japanese and the California Seventh-Day Adventist vegetarians, eat a lot of soy. In fact, the consumption of beans is one of the things that all the healthiest, longest-living ("Blue Zone") populations share in common.

INTRODUCTION

Soy. Mention its benefits in polite company and brace yourself for a barrage of questions and accusations directed at soy and all the products made from it.

Many people have been put off by claims made by a small group of people who have circulated numerous allegations about the safety and health benefits of soyfoods. Everyone seems to have an opinion. Concerned consumers often hear conflicting information in the media and particularly on the Internet, where sensationalism is the engine that can drive a story like wildfire.

Soy has been under attack in recent years. As interest in the health benefits of soy intensified in the research community, the popularity of soyfoods skyrocketed. Acceptance among consumers of what was heretofore a "hippie food" began to take hold. Innovative new soy products with mainstream appeal seemed to be entering the market almost weekly. As momentum heightened, soy products began penetrating mainstream America and another phenomenon emerged: "soy bashing."

Astounding claims began to surface about soy, creating anxiety and apprehension among many health-conscious consumers. Shocking assertions concerning infertility, sex organ abnormalities, Alzheimer's disease and thyroid problems have produced a great deal of confusion among consumers. The mighty soybean, which for centuries has held a special place at the table in the East, was now taking a beating in the West. A new cottage industry has emerged peopled by bean-bashers who have made condemning soyfoods an ongoing crusade.

Elevated blood cholesterol has been shown to increase the risk of cardiovascular heart disease, significantly. Soy protein has been credited with lowering blood cholesterol levels, and has been found to contribute to the increase of good cholesterol (HDL) and the lowering of triglyceride levels.[1 2 3] Numerous studies suggest that consuming soy can reduce the risk of cancer. Genestein, an isoflavone in soy, has been shown to inhibit angiogenesis, the growth of blood vessels that feed cancerous tumors.[4] Other benefits include maintaining a healthy weight and dramatically reducing the unpleasant side effects of menopause. The anti-soy rhetoric that has precipitated this debate vehemently deny these and other benefits associated with consuming soy and continue to fuel numerous allegations in need of rebuttal.

The Skinny on Soy addresses the continuing campaign to cast doubt on the safety of consuming soyfoods with peer-reviewed research and common sense. Soy has a long and venerable history spanning fifty centuries of Asian culture and has been the subject of more than seventy years of research in the scientific community.

If you are a health-conscious consumer interested in eating tasty, wholesome, and convenient plant-based meals more often, but have cut back or eliminated soy from the menu, this book is written for you. Perhaps you have been put off by the strident assertions that eating soy may be harmful. You may have heard that all of the positive benefits associated with eating soy are the result of a conspiracy to deceive people in order to make a profit.

Certainly, making a profit is the goal of any business; however, to suggest that there is a conspiracy afoot among the soyfoods business community to intentionally promote a harmful product to an unsuspecting public is a very serious charge. And one that has yet to be supported with evidence.

The contention that the soy industry is a powerful entity with major resources is very misleading. The US is certainly the largest producer and exporter of soybeans in the world. However, about 85 percent of the world's soybean crop is processed for animal feed. The soy industry that is being maligned here is, in reality, a very small component of the equation. Soy industry trade groups operate with budgets that are puny when compared to the meat and dairy industry, which makes up a large part of the major agribusiness interests.

It is also unrealistic to imply that the entire scientific community has somehow been manipulated for many decades in order to produce erroneous results. The nature of science is to challenge hypotheses, to question each outcome at every step of the way, and researchers in peer-reviewed studies do just that with plenty of oversight. Research scientists are investigators who question everything and reach conclusions that are almost always couched with the requisite "but further research needs to be done."

Conspiracy theories seem to take on a life of their own, no matter how specious and how many health-conscious consumers have been confused by the contradictory and often frightening headlines about soy. *The Skinny on Soy* speaks with top experts in the field to shed light on the issues and sort out the claims behind the mountain of anti-soy rhetoric. We take on the matter by reviewing and explaining the research in a way that makes sense—common sense.

There are millions of people who are deeply interested in the methods and findings of science. I am one of them.

Research is about asking questions and looking for answers. The twenty-four–hour news cycle has spawned an age where producers need to "feed the beast" with a never-ending supply of headline-grabbing news items and shock-inducing stories. It is in this atmosphere that the often-inconclusive statements about ongoing research are picked up, no matter how counterintuitive and are broadcast as "breaking news."

And how the news loves controversy! Abstracts do not provide the entire picture, and, quite often, intermediate studies are circulated beyond the community of researchers for which they have relevance and who are qualified to evaluate them.

On these pages, we will discuss the kinds of resources that have been used to impart a patina of credibility to some of the most dreadful charges ever leveled at a foodstuff. We hear directly from scientists who have been working with soy for decades. We define and consider the various methods and models used in research and how rumors, misinformation, and distortions allegedly based on science have played a role in soy bashing.

The Skinny on Soy examines the allegations at the source of these rumors, exploring the issues with research scientists, nutritionists, physicians, farmers, and manufacturers with years of firsthand experience with soy.

Sit back, pour yourself a tall glass of soymilk, and join me as we sort through the allegations and rumors that have cast a shadow on soy for well over a decade.

CHAPTER 1

Soyfoods on the Menu

Soy, a staple in Asian culture for centuries, became a niche market food source with countercultural appeal in the 1960s. Soy products such as tofu, soymilk, miso, and tempeh were often marketed by small companies to co-ops or small health food stores and stayed very much out of mainstream consciosness until the 1980s.[1] Soyfoods have come a long way since then.

In 1979, Vitasoy, a Hong Kong–based soymilk manufacturer, introduced their soymilk to the United States. Initial sales were so small that the product was sold store-to-store in San Francisco's Chinatown. Today, Vitasoy is one of the largest soyfood manufactures in the world, offering a broad range of soy products.

In the 1980s, soymilk began to gain market share as more people became aware of the health benefits of plant-based foods. In 1983, Eden Foods was one of the first companies to launch a national brand of shelf-stable soymilk packaged in aseptic boxes. Soymilk's popularity continued to increase among health conscious consumers well into the 1990s, even though it was still not generally available to consumers outside of health food stores.

In January of 1996, White Wave of Boulder, Colorado, introduced Silk soymilk in refrigerated half-gallon gable-top (typical milk carton) containers in supermarkets. It was a game-changer. Silk soymilk was the first breakthrough soy product in mainstream distribution and its innovative packaging won this refrigerated soymilk product a place in the dairy case at grocery stores across the country. By 2000, Silk soymilk was carried in twenty-four thousand supermarkets.

A *60 Minutes* segment with John Stossel in 2004 focused on the phenomenon when they used White Wave Silk soymilk in a blind taste test with cow milk with random mainstream consumers. Most people who tried it actually preferred Silk soymilk to cow milk, citing the taste and silky texture. Soymilk's huge success among mainstream consumers also invited the first in a series of attacks from competing market interests.

In 2012, *The Wall Street Journal* reported that the dairy industry was grappling with a vexing puzzle: how to revive cow milk sales. The authors noted that per-capita US milk consumption, which peaked around World War II, had fallen almost 30 percent since 1975.[2]

In a subsequent article in 2014, the *Journal* reported that cow milk consumption continued to decline as the sale of nondairy beverages continued to increase.[3]

An article in *The Globe and Mail of Canada*, "Milk sales continue to slide as diets, society shift away from dairy," noted that Canadian shoppers are increasingly skipping the dairy aisle at the grocery store.

Statistics Canada reported that sales of milk fell in June of 2015 by more than 3 percent from the same month a year earlier, marking the eighth consecutive monthly decline of what was once a staple of the Canadian diet.[4]

Sylvain Charlebois, a business professor at Ontario's University of Guelph, "It's been going down for at least 25 or 30 years."

In 2002, Dean Foods Company, a nearly ninety-year-old dairy giant embraced the nondairy beverage trend and acquired White Wave, Inc., makers of Silk soymilk. However, in 2013 Dean Foods completed the spinoff of The White Wave Foods Company divesting itself of the remaining shares.

Why would they do that? While White Wave's profit and sales climbed as US consumers embraced plant-based milks, Dean Foods churned out losses on falling domestic demand and higher costs for raw milk. Today, White Wave's revenues are just one-third of Dean Foods, but its market value is more than three times its former parent.

Getting milk drinkers back in the barn won't be easy. "It is going to be tough to buck the trend of declining consumption of [cow] milk in the U.S.," says Ryan Oksenhendler, an analyst with Arlon Group LLC, a New York–based fund manager that owns White Wave shares. He says shoppers quitting cow milk and embracing soy and other nondairy milks are feeding White Wave's gains.

Nancy Chapman, executive director of the Soyfoods Association of America told *USA Today* that soymilk sales in 1996 were

$124 million and topped one billion dollars by 2008. Cow milk retail sales were $12.3 billion in 2008, meaning soymilk has been making a big dent in dairy market share. The dairy industry has clearly been threatened by ongoing market trends, which have not reversed in almost twenty years.

In February of 2000, the National Milk Producers Federation (NMPF) filed a complaint with the FDA seeking to restrict the term "milk" on product labeling to "mammalian lacteal secretions."

In 2010, the NMPF, for the second time in ten years, petitioned the Food and Drug Administration asking that the term "milk" be reserved for cow milk. They conceded that it is acceptable to use the word for goat, sheep, or water buffalo milk.

According to FDA spokesperson Siobhan DeLancey, "We evaluate all these communications, but we plan our actions based on what will make the most impact for the public health." However, the FDA has not responded to either complaint in more than fifteen years and has yet to react.

Cow Milk Does a Body Good?

Sixty percent of the world's population is unable to digest cow milk and scientists say that lactose intolerance is actually a misnomer, as it implies an abnormal condition.

The ability to digest milk from any species in adult populations is not normal, and the level of intolerance breaks down as follows: 5 percent of Asians, 25 percent of people of African and Caribbean descent, 50 percent of Mediterranean people, and 90 percent of Northern Europeans (with Sweden having one of the world's highest percentages of intolerance). The ability to digest milk is actually a weird genetic adaptation that would more accurately be described as "lactose persistence."

The dairy industry spends billions of dollars to convince the public that cow milk is a beneficial commodity that provides a unique set of essential nutrients. They drill slogans through the airways, in print and online like the ubiquitous milk mustache "Got

Milk?" campaign to convince consumers that cow milk is a healthy beverage and that their life would be incomplete without it.

Ads that promote dairy products are selling one of two overall concepts: (1) "To be strong, active and healthy, drink low fat cow's milk," or (2) "Indulge in high fat dairy products . . . a guilty pleasure."

USDA-managed programs alone spend $550 million annually to bombard Americans with a barrage of advertising urging us to buy more animal foods.

These days, it's not just about soymilk. The US retail soyfoods industry totaled $4.5 billion in 2013, up from $1 billion in 1997. Soyatech is an agricultural event, media and consulting firm; they reported that soyfoods sales in 1980 were just $300 million.

Soy burgers, sausages, bacon, beverages, frozen desserts, snacks, cookies, crackers, pasta, and more are easy to find in big cities and small towns across America. Soyfoods are on the menu just about everywhere. You can have a soy burger for lunch, soy latte on a coffee break, and gourmet tofu dishes at some of America's best restaurants.

Whether for health or for philosophical reasons, people who buy soyfoods believe they bring a lot to the table. In addition to the highly adaptable texture and variety that soyfoods like tempeh, tofu, and the burgeoning category of meatless alternatives contribute to the menu, numerous scientific studies have reported many health benefits associated with soy.

Frequent references are made to peer-reviewed research throughout this book. Scientific knowledge is cumulative and builds on itself and peer-reviewed work meets certain standards of scientific quality. Chapter 14 offers the opportunity to gain an understanding of the methods and criteria used in various types of scientific studies.

Soybeans have been shown to contain an impressive array of nutrients and sub-nutrients that would dwarf most other foods, plant-based or otherwise. For decades, scientists have produced numerous studies demonstrating that soybeans are the source of complete high quality protein, complex carbohydrates, soluble and insoluble fiber, and unique phytochemicals credited with the prevention of many chronic diseases.

It is widely accepted that soybeans contain bioavailable calcium, iron, zinc, fiber, and several B vitamins, and a source of omega-3 fatty acids thought to reduce risk of heart disease and high blood pressure. With so much going for it why is this awesome bean taking a bashing?

CHAPTER 2

Fact, Fiction, and Fallacy

"A lie gets halfway around the world before the truth has a chance to get its pants on."

—Sir Winston Churchill (1874–1965)

Soy has become a superstar among healthy alternatives. As is often the case in the media, a person or product will enjoy great coverage, achieving celebrity status only to be knocked down when the press pendulum swings the other way.

Perhaps, after so much good news about the numerous health benefits associated with a diet rich in soy, from menopause relief to cancer prevention, a backlash was almost to be expected. In recent years, a virtual avalanche of sensationalistic rumors and accusations has stalked soyfoods.

Soy has become the target of some rather serious allegations purportedly refuting decades of research regarding the health benefits associated with consuming soy. Rumors have been circulating about substances in soy that may be harmful, causing everything from infertility and cancer to homosexuality. I am a health writer and soyfoods expert and have often been asked to respond to questions from concerned consumers distressed by these frightful claims.

The Internet can be an amazing resource for the exchange of ideas and information. For students, writers, and authors doing research or consumers searching for food and health information, the Internet delivers a world of data to the desktop. However, this amazing technological behemoth can easily become an instrument of misinformation and controversy.

This egalitarian resource can often generate a great deal of inaccurate information and has become the home base for a campaign to cast serious doubt on the wisdom of consuming soy. It is on the Internet that you will find sensationalistic claims based on half-truths and junk science on a variety of topics, with soy at the top of the list.

When these allegations are scrutinized, what is most surprising is that they have originated with a small contingent of very vocal activists. The first wave of queries came to me following a fervently anti-soy article, "Tragedy and Hype," by Sally Fallon and Mary G. Enig.[1] The authors portray soybeans and the myriad of soyfoods made from them as poisonous and unfit for human consumption.

According to Fallon and Enig, soybeans contain substances that cause hypothyroidism, cancer, and sex organ abnormalities—from young boys growing breasts to stunted growth.

"Tragedy and Hype" makes a number of rather grave charges that indict the entire soyfoods industry, alleging a concerted conspiracy to conceal damaging information.

Fallon and Enig declare that those who will be held legally responsible for deliberately manipulating the public for financial gain "include merchants, manufacturers, scientists, publicists, bureaucrats, former bond financiers, food writers, vitamin companies and retail stores."

Food writers? Vitamin companies and retail stores? Bond financiers? The litany of dangerous health hazards these authors attribute to consuming soyfoods would put just about every area of the body at risk of dire consequences.

The Internet is an echo chamber and allegations, recycled and rebroadcast over the years, leading to a number of articles, posts, books, and websites with similar commentary referencing the authors. As a result, many people now have serious reservations about the safety of consuming soy. Note the many disclaimers on product labels prominently declaring "soy-free," even though manufacturers may know soy to be a nutritious and beneficial ingredient; they are forced to acquiesce to the fears of consumers.

When evaluating information, it is important to consider the source. The most strident anti-soy naysayers are those associated with the Weston A. Price Foundation (WAPF). The WAPF was established decades after Dr. Price's death in 1948 and represents meat and dairy farm community chapters around the country. Their financial and philosophical interests are in promoting the consumption of meat and dairy products, particularly raw milk and in villainizing soy and vegetarianism.

The Price-Pottenger Nutrition Foundation of La Mesa, California, was founded in 1952 as the Santa Barbara Medical Research Foundation, became the Weston Price Memorial Foundation in 1965, and adopted its current name in 1969.

The Weston A. Price Foundation was named in honor of an early twentieth-century dentist and author of Nutrition and Physical Degeneration who traveled to remote regions of the world in the 1930s.

Dr. Price was a dentist who studied the diets of indigenous tribes to determine the link between diet and dental health. The WAPF advocates a diet high in beef, pork, and other flesh foods with generous amounts of butter and raw milk, none of which have even a passing relationship with Dr. Price's work.

In 1934, Price wrote to his nieces and nephews, recommending a diet based on whole foods: "The basic foods should be the entire grains such as whole wheat, rye or oats, whole wheat and rye breads, wheat and oat cereals, oat-cake, dairy products, including milk and cheese, which should be used liberally and marine foods."

It is interesting to note that the Weston A. Price Foundation promotes the consumption of beef, pork, and other meat products high in fat, saturated fat, and cholesterol with fervor and disparage anyone who consumes a diet rich in whole grains.

Dr. Price observed that many native cultures were very healthy while eating lacto-vegetarian or basically vegan diets with a small amount of fish. Sally Fallon, the foundation's president never misses an opportunity to denounce vegetarianism, with article headings such as "Twenty-Two Reasons Not to Go Vegetarian," "Myths & Truths About Vegetarianism," and "Nutrient Deficiencies on a Vegetarian Diet," which try to contradict peer reviewed evidence suggesting the reduced risk of many diseases and increased health, quality of life, and longevity associated with the vegetarian and vegan lifestyle.

Sally Fallon's book, *Nourishing Traditions: The Cookbook that Challenges Politically Correct Nutrition and the Diet Dictocrats*[2] offers advice that is essentially counter-intuitive and employs mocking and manipulative language toward anyone who would disagree.

Written with Mary Enig, this book lists more than two hundred references; however, many are obsolete and the authors almost exclusively reference their own writings or those of other WAPF authors. Very few of these references were published in peer-reviewed journals and they were generally misrepresented.

The regimen recommended by the WAPF contains unhealthy levels of fat, saturated fat, and cholesterol. Coronary Heart Disease (CHD) occurs when arteries become clogged with plaque, a condition called atherosclerosis.

These plaques, which consist of fat and cholesterol build up on the interior of the arteries reduce the flow of oxygen-rich blood to the heart. A diet high in animal products introduces unhealthy levels of fat and cholesterol and increases the risk of CHD.[3 4]

Public health officials and the state departments of agriculture strongly discourage the drinking of raw milk on the grounds that it can be a dangerous, germ-ridden beverage that is especially hazardous to children because of their immature immune systems.

The immune system is made up of a network of cells, tissues, and organs that work together to protect the body. It is the body's defense against infectious organisms and other substances that invade the body and can cause disease.

Raw cow milk has been a source of bovine tuberculosis and drinking milk that has not been pasteurized increases the risk of illness that may require hospitalization, as well as the risk of transferring other maladies such as mad cow disease.[5 6]

Since 1987, the FDA has required pasteurization of almost all packaged milk products with the exception of some kinds of aged cheese; however, dairy-borne disease is still very much a threat.

The agency noted that unpasteurized milk products may contain a variety of infectious bacteria such as Salmonella and E. coli, which can be the cause illness or even death.

Campylobacter is a bacterium that causes gastrointestinal symptoms that include diarrhea that may be bloody, accompanied by nausea, vomiting, cramping, and fever, which develops within two to five days of exposure. These symptoms typically last about a week; however, it is possible for an individual to be infected with campylobacter and not exhibit symptoms.

According to the Centers for Disease Control (CDC), in individuals with compromised immune systems, Campylobacter can spread to the bloodstream and cause a serious life-threatening infection.

The Foodborne Diseases Active Surveillance Network (Food -Net) indicates that there are about fourteen cases of campylobacteriosis diagnosed per one hundred thousand persons annually. Many more cases go undiagnosed or unreported and campylobacteriosis is estimated to affect over 1.3 million people every year.[7]

Although campylobacter infection does not commonly cause death, it has been estimated that approximately seventy-six people with campylobacter infections die each year.

In March of 2010, an outbreak of campylobacteriosis prompted the Food and Drug Administration (FDA) and several mid-western state health agencies to warn consumers against drinking raw milk.

There were recurrent outbreaks of campylobacter infections associated with raw milk in Pennsylvania that came from a dairy that was certified by the Pennsylvania Department of Agriculture to sell raw milk. During January and February of 2012, the dairy was identified as the source of a multi-state outbreak of campylobacteriosis[8] and then again in April and May of 2013.[9]

In September of 2014, officials with the Wisconsin Department of Health Services reported that unpasteurized (raw) milk served at a potluck team meal is the likely cause of a campylobacter outbreak that sickened at least twenty-two members of the Durand High School football team.[10]

In March 2015, California public health investigators learned of six northern California residents who were diagnosed with campylobacter as a result of consuming raw milk produced by a local farm. The state investigators isolated campylobacter in multiple bottles of the raw milk from that farm.[11]

Some point out that the WAPF does help to raise awareness about the dangers of processed food. However, much of what they espouse is completely out of step with accepted medical principles and often challenges common sense.

The counterintuitive health advice does not end with artery-clogging animal foods and risky raw milk. The WAPF continually endorses many inaccurate and outdated dietary recommendations that do not meet the nutritional standards of modern science and are potentially dangerous.

Unusual Dietary Recommendations of the WAPF:

- The feeding of sea salt to infants and babies
- Homemade raw milk formula with whey, lactose and oils
- Egg yolks for infants as early as four months
- Cod liver oil for infants from four months of age
- Introducing chicken livers and beef broth at six months of age
- Limiting fruits and vegetables in children's diets
- Homemade stock made from animal bones
- Butter and shellfish and castor oil compress for constipation
- Liberal use of animal fats such as lard, tallow, and butter
- Full-fat dairy products, preferably raw
- Egg yolks, cream, coconut, and palm and palm kernel oil
- Adding poached animal brains to other ground meats

Since the 1950s, about twenty thousand articles have been written documenting the link between a diet high in saturated fat and low in vegetables, beans, fruit, and other plant-based foods and the increased risk of heart disease and cancer. Coronary heart disease is the leading cause of death in the world and there are thousands of research scientists who would take the recommendations of the Weston Price Foundation to task.

Quackwatch describes the Foundation as promoting questionable dietary strategies: "Its (Weston Price Foundation) newsletter, book catalog, and information service promote food faddism, megavitamin therapy, homeopathy, chelation therapy, and many other dubious practices."

According to best-selling author and health advocate John Robbins:

> Weston Price never once mentioned the words soy, soybean, tofu, or soymilk in his five hundred–page opus and spoke quite positively about lentils and other legumes, yet the foundation has taken it upon itself to be vehemently and aggressively anti-soy.
>
> Regrettably, those currently running the Weston A. Price Foundation seem to be oblivious to the spirit of compassion, which motivated the work of the man under whose name they act.
>
> Sadly, they are not just intolerant of people who eat or think differently than the way they advocate; they frequently demean and condemn those with whom they disagree. There is a nastiness, a mean-spiritedness to their activities that is not worthy of the man in whose footsteps they presume to follow.

Robbins continues:

> The Weston A. Price Foundation exudes an attitude of "you're either with us or you're against us" that is reminiscent of the dark side of cults.
>
> Those authors and researchers who the foundation disagrees with are caustically mocked. If those authors happen to subscribe to the findings of modern nutritional science, they are condemned for being politically correct. Reputable scientists who dare sug-

> gest that saturated fat contributes to heart disease are denounced for being as PC as PC can be—and totally ignorant.

The forces promoting the anti-soy information that has garnered so much attention make no attempt to mask their anti-vegetarian agenda. They quote many obscure animal studies and often reference a single study with inconclusive results that was never duplicated.

Their articles and websites advance the idea that saturated fat and dietary cholesterol are not a cause of heart disease, and that they do not interfere with blood vessel function.

That's quite a news flash for millions of people who have undergone angioplasty and bypass surgery. The soy antagonists contend that cholesterol, which is found exclusively in animal products, and saturated fat that is only found in any significant amount in tropical oils in the plant kingdom, are actually healthful dietary components.

To suggest that arteries clogged with plaque made up of saturated fat and cholesterol do not reduce the flow of blood to critical organs such as the heart and the brain is illogical and unsound.

To suggest that dietary fat and cholesterol do not contribute mightily to diseases such as atherosclerosis and Alzheimer's disease is an opinion that does not have the support of modern medicine.

A quote from researchers of one study published in the *Journal of the American College of Nutrition*:

> During the past several decades, reduction in fat intake has been the main focus of national dietary recommendations to decrease risk of coronary heart disease (CHD).[12]
>
> Metabolic studies have long established that the type of fat, but not total amount of fat, predicts serum cholesterol levels. In addition, results from ep-

> idemiologic studies and controlled clinical trials have indicated that replacing saturated fat with unsaturated fat is more effective in lowering risk of CHD than simply reducing total fat consumption.

Chris Masterjohn works for the WAPF and has written a number of articles, such as, "Meat, Organs, Bones and Skin"[13] and "Beyond Cholesterol"[14] extolling the benefits of a high fat, meat-based diet loaded with saturated fat and cholesterol.[15]

He, along with a few other WAPF followers, has embarked on a noisy and rather nasty campaign to discredit *The China Study*, which was published in 2006 by T. Colin Campbell, PhD who authored the book with his son, Thomas M. Campbell II, MD.[16]

The China Study examines the relationship between consuming animal products and the onset of many chronic diseases, such as cancers of the breast, prostate, and large bowel, coronary heart disease, diabetes, and obesity, as well as osteoporosis, degenerative brain disease, macular degeneration, and autoimmune disease.

The New York Times called the book, which draws on The China Project conducted by Cornell University, Oxford University and the Chinese Academy of Preventive Medicine over the course of twenty years,[17 18] "The Grand Prix of epidemiology."

The China Project was conducted in China where genetically similar populations tend to live in the same way, in the same place, and eat the same foods for their entire lives.

The China Project analyzed a survey of death rates for 12 kinds of cancer in over 2,400 counties and 880 million people and studied the relationship between mortality rates and dietary, lifestyle and environmental factors in 65 counties in China, which were mostly rural.

What was Dr. Campbell's conclusion? A whole foods, plant-based diet is the optimal nutritional approach to reduce the risk of chronic illness and disease. The extraordinary success of this book and Dr. Campbell's voluminous and well-regarded body of work has made him a target for these anti-soy crusaders, who are vehemently anti-vegetarian.

Another source of strident anti-soy invective is The SoyonlineService, a website operated by Richard James and his wife, Valerie, who are New Zealand dairy farmers and bird breeders.

Their widely broadcast claims that the phytoestrogens in soy have deleterious effects began, following the devastating demise of a number of birds in their care.

Apparently, in 1992 this New Zealand couple fed their parrots a variety of bird food, which contained soy and corn flour. To their horror, the birds began to sicken and die after suffering multiple organ failure.

They attributed the illness and subsequent demise of their birds to soy present in the bird feed, even though it contained other ingredients.

A most reasonable explanation might very well be that the corn in the feed was infected with a common fungal pathogen, which is well known to cause the kind death and disease suffered by their birds.

Mold is a term used to describe the various forms of fungus that can contaminate crops and animal feed. Mycotoxins are the highly toxic byproduct of mold growth. *Myco* means fungus and *toxin* means poison. Mycotoxins can cause a broad spectrum of acute and chronic disease in both humans and animals.[19]

Fungal toxins, such as zearalenone or aflatoxin, are molds that are common in crops such as corn and wheat and are caused by poor storage or growing conditions. A sampling of corn in Argentina in 1995 showed the universal presence of zearalenone.[20]

When certain types of fungus grow on food, they produce small amounts of mycotoxins. Some of these fungi (primarily Aspergillus flavus) produce the very lethal mycotoxins called aflatoxins. Aflatoxins are remarkably potent, often causing disease even when ingested in minute amounts.

Mold that is both airborne or found growing on food is dangerous to parrots. Aspergillus mold can cause the deadly disease, aspergillosis, and is found growing on foods that have been handled or stored improperly. Aspergillus mold is particularly common on grains such as corn.

Although any variety of mycotoxin can cause health problems, aflatoxins are the one that is most likely to appear in bird feed. Birds are well known to be extra sensitive to aflatoxin exposure and the resulting liver damage.

When consumed over time, even small amounts of this fungal contaminant can weaken liver function. The liver has a number of functions such as detoxification, the workings of the immune system, and digestion. Therefore, aflatoxin can cause liver damage and disease and also lead to other chronic disorders.

Corn is the major crop affected by Zearalenone and when feed grain contaminated with this fungus is ingested, the result can be the onset of a wide variety of reproductive problems.

It is a matter of some concern that these poisons are completely heat stable and can remain on the food indefinitely. As soon as the fungus grows on the plant and produces the aflatoxin, neither cooking nor freezing will destroy it.

Dick James contends that soy is at the root of their bird breeding issues. However, if soy were the cause of immune system shutdown and multiple organ failure, why haven't we heard of any other such claims?

Surely many more breeders would be reporting health issues with such a dire consequences. To attribute the illnesses and mortality their flock experienced to a single ingredient not uncommon in bird feed just doesn't seem logical. It is more feasible to suggest that the parrots in New Zealand were fed contaminated bird feed, unbeknownst to their breeders, Dick and Valerie James.

Further allegations emerged from the SoyOnLineService attributing everything from thyroid disease to the rise in homosexuality to the phytoestrogens in soy, dubbing them potent chemical toxins.

The SoyOnlineService has mounted an Internet-based crusade against soy products. The first paragraph on their website, soyonline.com, is a wholesale indictment of the soy industry, characterizing the entire industry as "lying marketers" who hide the truth about soy.

The SoyOnlineService website makes absurd unsupported statements such as, "Soy was never a staple in Asia," and that "Asians eat very little soy." These anti-soy activists have launched a crusade to have soy formula removed from the marketplace and, along with the WAPF, have been threatening a class action lawsuit in New Zealand and Australia since 2007.

We sort through these and other allegations here; however, it should be noted that the phytoestrogens in soy are also found in other plant foods, notably garbanzo beans, sweet potatoes, red clover, taro, and millet.

I cannot move on without pointing out that homosexuality has been documented from the beginning of recorded history among humans and animals. There is no correlation between homosexuality and whether a culture does or does not eat soyfoods, and there is no indication that the incidence of homosexuality is on the rise. If anything has changed it is that society is increasingly accepting of homosexuals and they can be more open.

Fallon and Enig call soy infant formula "birth control pills for babies," and have promoted Dick and Valerie James and their anti-soy formula campaign extensively.

Soy infant formula has been on the market for decades and yet no pattern of abnormalities such as those described by the James' SoyOnlineService have emerged among the millions of adults who were nourished with soy formula as infants.

Thomas Badger, PhD Director, Arkansas Children's Nutrition Center, Arkansas Children's Hospital Research Institute explains:

Modern soy formula has been around for probably thirty years or more and in that period of time there have been over twenty-five million kids who have grown up on soy formula and there are no verified reports in valid peer reviewed journals with adverse effects of soy formula. So that's got to tell you a lot—I mean these kids grow up to be as normal as any other kids as adults.

Other players who have emerged among the anti-soy contingent include Joseph Mercola, D.O. He is an osteopathic physician and online marketer with a newsletter and a wide-ranging line of products, from supplements, shower filters, and tanning oil to saunas and bamboo toilet paper.

Dr. Mercola claims that consuming pasteurized milk instead of raw milk is a cause of autism and that coconut oil is the well-kept secret to weight loss, detoxification, and reversal of heart disease, and that it kills viruses!

Dr. Mercola, who is a very vocal critic of soy, has been reprimanded by the US Food and Drug Administration (FDA) and has received three warning letters for making unsubstantiated claims regarding products such as coconut oil, chlorella, and diagnostic cancer screening tools.[21 22 23]

Dr. Mercola and the WAPF preach that saturated fat is good for you and quote articles with a modicum of dissent while ignoring thousands of well-documented, peer-reviewed studies to the contrary. The writers quote each other and ignore reputable modern-day research. Worse, they often distort what has been stated in legitimate studies in order to shore up assertions such as "saturated fat is healthy and not related to heart disease."

Dr. Mercola has a website devoted to soy with headlines such as, "Soy: This 'Miracle Health Food' Has Been Linked to Brain Damage and Breast Cancer," citing "hard evidence" for a litany of maladies. And whom does he cite? WAPF author, Kaala Daniels.

If the avalanche of allegations weren't so serious, then claims like boys growing breasts and soy causing infertility would be hilarious and all of Asia would be long overdue for a meltdown.

The greater tragedy is that all of the hype and hysteria has frightened consumers and denied them the health benefits of including soyfoods in their diet.

CHAPTER 3

Rumor: Soy Is an Incomplete Protein

Old myths die hard, and there are those who would have you believe that a plant-based meal is somehow lacking, particularly in protein. Some reach back in time and insist that one must follow the "food combining" strategy promoted in the 1971 best-selling book *Diet for a Small Planet*. (The author, Frances Moore Lappe, has since reversed her position in the 1981 edition.)

Soyfoods have sustained one of the most complex and venerable cultures in recorded history and it is no secret that Asian people often enjoy vibrant health and longevity into old age. The Chinese have been known to refer to tofu as "meat without bones" and what the ancients knew intuitively about soy has been documented by modern science.

Soybeans contain a higher level of protein than other beans. They are relatively low in carbohydrates and get 35 to 38 percent of calories from protein, whereas other legumes contain 20 to 30 percent protein.

Soy protein is a complete high quality protein comparable to meat, milk, and eggs, but without the unhealthy cholesterol and saturated fat associated with animal products. That would make soy an exceptional source of dietary protein and an excellent choice in terms of displacing foods that may be high in protein but offer less overall nutritional value.

The World Health Organization (WHO) addressed the quality of soy protein and whether or not it supplies all of the necessary amino acids twenty-five years ago. A 1991 Food and Agriculture Report (FAO) identified soy as a complete protein that meets all of the essential amino acid requirements of humans.[1]

The 1988 American Dietetic Association Position Paper discredited "Protein Combining," stating, "Adequate amounts of amino acids will be obtained if a varied vegan diet, containing unrefined grains, legumes, seeds, nuts, and vegetables is eaten on a daily basis."[2]

For at least five thousand years, soy has been a dietary staple and the primary source of complete, high quality protein for millions worldwide. So, where did this flawed information originate?

It came from an outdated method of evaluating protein requirements, the Protein Efficiency Ratio (PER), which based the protein quality for humans on the growth of young rats, whose amino acid requirements are vastly different from humans.

From 1919 until recent years the PER method had been widely used for evaluating the quality of protein in food. In 1993 the FDA adopted the PDCAAS method, which is based on human requirements for amino acid and is more appropriate than a method based on the amino acid needs of other species.

James Anderson, MD is a Professor of Medicine and Clinical Nutrition at the University of Kentucky and founded the HCF Nutrition Research Foundation in 1979. Dr. Anderson has been working on soy for over twenty years and has published twenty-five papers on the subject. Dr. Anderson was the lead investigator on the groundbreaking meta-analysis that predated the 1999 FDA approval of the soy protein and heart health claims.[3]

According to Dr. Anderson, soyfoods are the healthiest foods that you can put on the table. "I feel strongly that they are really health promoting with respect to heart disease, diabetes, bone health and kidney function."

He says that soy protein is the component that provides benefits for the heart:

> I've looked carefully at cholesterol. I am sure soy protein lowers blood cholesterol. The data is very clear to me. Many of the heart health elements are related to the protein, not to the phytoestrogens.
>
> The proteins are what lower blood cholesterol. It's complicated but the peptides (the short amino acid chains) are really the active ingredients for lowering blood cholesterol. I've looked at blood pressure doing a meta-analysis on soy protein lowering blood pressure and it is the peptides that are doing that.

Americans have long ago bought into the "protein myth," a belief that protein exists only in animal products and that one needs copious amounts of protein in order to thrive.

Nothing could be further from the truth. Certainly, protein is an important component of a healthy diet and it is filling and the most satiating of all the nutrients. We need protein to build cells and repair damaged tissues in the muscles, bones, skin, cartilage, and blood. Protein also plays a role in making the enzymes and hormones that keep the body functioning.

So, yes it is important; however, protein is widely available apart from animal products. You don't need to eat muscle in order to make muscle cells in your body.

Additionally, the Western diet is overloaded with protein. Most people in the US consume at almost twice the amount of protein they need or is healthy for them. Excessive dietary protein adds more nitrogen than the body needs and is excreted placing a strain on the kidneys and can lead to kidney damage.[4]

Another unhealthy side effect of protein overload is that sulfur-based animal protein leaches calcium from the bones on it's way out of the body, increasing the risk of osteoporosis.[5]

If you are interested in the benefits of a vegetable-based diet and are concerned about consuming enough protein, there is an easy formula that will calculate just how much protein an individual requires. Simply determine your ideal weight, divide that number by kilograms (2.2) and then multiply that number (your weight in kilograms) by 0.8.[6]

For instance, an ideal weight of 150 pounds divided by 2.2 = 68 X 0.8 = 54.5 grams. That means a 150-pound individual requires less than 55 grams of protein per day.

The National Health and Nutrition Examination Survey (NHANES) conducted for the National Center for Health Statistics found that the average American male consumes 102 grams of protein per day, while the average female consumes about 70 grams. That is almost twice the recommended daily intake, primarily in the form of animal protein, which also adds unnecessary amounts of fat, saturated fat and cholesterol.

It is important to note that the recommended daily requirement (RDA) for protein is calculated at two standard deviations *above* the mean requirement, which becomes the public health recommendation. The RDA represents a population goal. That means the RDA for protein 50 to 55 grams a day for women and 65 grams a day for men contains a "safety margin" designed to meet the needs of 97.5 percent of the population. Simply put, the RDA contains a little extra cushion.

Neal Barnard, MD, is a nutrition researcher and founder and president of Physicians Committee for Responsible Medicine in Washington, DC. He says, "I think soy products are safe and they have certain health advantages, particularly related to cancer prevention. That's true both for women and for men."

Dr. Barnard doesn't feel that soyfoods are an absolute essential; however, he believes that they do present some clear-cut advantages, not the least of which is what they can replace. Dr. Barnard says:

> It's very important to see what soy can replace. First off, a person eating a veggie burger made of soy instead of a hamburger is not only getting the advantages of soy, they are skipping all the risks of a meat burger. If you are eating soy cheese rather than regular cheese, you are going to be doing much better with regard to saturated fats.

When people hear news reports about the health supporting properties of certain foods—be it carrots, broccoli, walnuts, flaxseeds, or soy—they are keen to add these foods or supplements to their diet. The Western approach to health care, which treats symptoms rather than the source, has created a mainstream mindset on a quest for the "magic bullet."

I asked Dr. Barnard about the effects of adding soy to a diet that still contains unhealthy amounts of fat, saturated fat, and cholesterol.

He said "Many, many researchers have shown the value of moving away from animal products to a plant-centered diet and so the more we get away from meaty, cheesy, fatty products and more towards plant products including soy, we're going to do so much better."

These days, losing weight and living a healthier lifestyle is all over the media and pretty much dominates the collective conversation. It is unfortunate that so much misinformation has been disseminated about soy and has alarmed so many who might otherwise have enjoyed its benefits.

Dr. Barnard explains:

> I think many people are not up to speed on what the latest research has shown. It's certainly true that when we talk about soy products, the commercial interests try to weigh in and industries that feel threatened by the value of soy sometimes come out and make criticisms that end up not standing the test of time, and they end up frightening people along the way.
>
> Luckily, when you look at what is happening in stores, it's pretty obvious that the public is recognizing that there is value to soy products. Not only are they very tasty, but they see the health benefits and the benefits of replacing the unhealthier food and that is certainly all to the good.

According to Dr. Anderson:

> The only adverse effects of soy relate to the occasional soy allergy, which is less common than peanut allergy, and also there's intolerance. When people start eating soy protein and soyfoods, it may affect their

gastro-intestinal (GI) tract and sometimes there are the acclimations to the taste.

I wrote a detailed review of this. I had to look at this rigorously several years ago and I have followed the literature since then. There are really no health disadvantages to soy except the occasional soy allergy. I really believe that.

In October of 1999, the US Food and Drug Administration (FDA) authorized the use of health claims about the role of soy protein in reducing the risk of coronary heart disease (CHD) on the labeling of foods containing soy protein.

This final rule was based on the FDA's conclusion that foods containing soy protein included in a diet low in saturated fat and cholesterol may reduce the risk of CHD by lowering blood cholesterol levels.

Christopher D. Gardner, PhD, is the Director of Nutrition Studies at Stanford Prevention Research Center and an Associate Professor of Medicine at Stanford University. He is a staunch supporter of a plant-based whole foods diet rich in traditional soy products. Dr. Gardner explains:

> The FDA health claim that came out of all of the soy research was just about protein. In the amino acid profile of soy there is more arginine proportional to other amino acids than in other foods and arginine is potentially the heart-healthy component in soy protein.
>
> So, what we ended up with is a lot of soy protein added to junk food in order to make a health claim. And that's a sad reflection on the way we do our science and our food marketing.

I asked Dr. Gardner about the difference between the traditional soyfoods such as tempeh, edamame, and tofu and the popular new soy products such a veggie burgers, soy hot dogs, and ice cream.

What kinds of soyfoods does he like to include in the menu? What does he think about the products that are created with Isolated Soy Protein (ISP) and Textured Vegetable Protein (TVP)? Dr. Gardner, a vegetarian, had this to say: "I don't actually eat whole soybeans very much in soups or other dishes; I tend to use other beans, but I certainly could."

As with all plant foods, Dr. Gardener prefers whole soyfoods to the more processed soy products on the market, and likes to refer to them as junk food.

My kids love edamame. They take it to school in their lunchboxes all the time. Edamame is the whole bean. Soymilk is a fraction of the whole bean. When making soymilk, you have removed the fiber and lost some folic acid and some other things in the process. When you make tofu out of the soymilk, it is only a fraction of the soymilk and then by the time you get to the soy hot dogs and the soy ice cream, it's junk!

In making these kind of soy products, the process involves isolating the protein out of the soybean and combining it with additives and sugars and all kinds of things to make a food that Americans are familiar with, like a hot dog, and I don't really think a soy hot dog is healthful, in and of itself.

What about the role of soy in replacing unhealthy animal products, which make up the majority of the Western diet? Isn't replacing a traditional hamburger or hot dog with a soy-based alternative a far better nutritional choice?

Dr. Gardner says, "I think there's potential for soy to be a much more valuable and healthful food than it has become. So your initial point that maybe you are displacing a traditional hot dog, which is less healthy, there is definitely merit in that."

While Dr. Gardner agrees that replacing animal products that are loaded with fat and cholesterol has merit, he believes you are not getting kind the healthful benefits of a more traditional soyfood.

If someone wants the health benefits of soy, they should eat tempeh. Tempeh is the whole bean. A lot of us are very uncomfortable with change, and the soy junk foods (the various analogues that are made with ISPs and TVPs) are practical.

I will completely admit that this is practical for people who were raised on a meat-based diet and where there is this barrier of familiarity for them to overcome.

On the other hand, tempeh is a better option than some of these other choices, because tempeh is made from the whole bean. If you are going to slice up veggie hot dogs to put them in some dish as an alternative to a meat product, you would be way better off cutting up squares of tempeh and adding them to a familiar recipe.

Tempeh is a traditional soyfood fermented from whole soybeans that originated in Indonesia. Tempeh is high in protein, fiber, calcium, B vitamins, and iron, and it has a tender, chewy texture that most meat-eaters appreciate.

This delicious soyfood should be cubed and steamed for fifteen minutes before adding to a recipe; this will tenderize it and allow the tempeh to more readily absorb sauces and marinades. You can freeze raw tempeh for three months and once it has been steamed tempeh will keep covered in the refrigerator for up to five days, ready to be incorporated into your favorite recipe.

Dr. Gardner says, "Soy contains folic acid and it certainly has a great amino acid profile; it is a rich source of protein and if you eat the whole bean, you get the fiber."

Tofu is still a rich source of high quality protein, and what about tofu coagulated with calcium sulfate? According to Dr. Gardner:

> If you eat tofu and if they have added calcium to make it curdle, it will become a better source of calcium than say, tempeh. If you were to take a whole food approach, soy could easily be part of a really healthy diet and strategically soyfoods could help people to

transition from a less plant based diet to a more plant based diet, but soy isn't essential. You could also do it without soy.

An excellent case can be made for including soyfoods in the diet. Soy adds variety and high quality protein as well as other important nutrients and sub-nutrients to the diet and can play an important role in replacing heart-heavy animal products that dominate Western cookery. The myriad of innovative soy products, from veggie burgers to ice cream, have become commonplace have been instrumental in increasing the level of acceptance among mainstream consumers.

Dr. Gardner raises concerns that have been voiced by others regarding the nutritional value of these hybrid products, which are often the result of an ingenious blend of soy protein and wheat gluten. Should soy burgers and chicken-less McNuggets be considered "junk food"?

I asked best-selling author of *The China Study* and respected nutrition researcher, Dr. T. Colin Campbell his opinion regarding whether there is a huge disparity between traditional soyfoods and what some refer to as soy "junk food."

I consider those types of questions to be quite trivial and I don't mean to be pretentious. The reason I say that is as follows: The main relationship between food and health is best indicated by the total nutrient composition of these foods, collectively speaking. That is the big effect.

Now, one begins to recognize that there are many secondary questions that have to do with how does this food compare with that food or how does this form of this particular food compare to another form of that same food. Those questions, in terms of the effects that they have compared to the big effects, for instance plant

> foods versus animal foods or whole plant foods versus processed foods, are the really big questions.
>
> The questions as to whether soy products are better than the whole soybean? Yes, I am sure there are some differences and I am familiar with some of these claims.
>
> There is the question about whether the phytoestrogens present in soy products are important to consider and prepared soy products compared to tofu versus whole soy. And the related question concerning the fact that other legumes have the same properties that soybeans do, for the most part. It's more interesting to talk about legumes as a class rather than one individual food product.

Dr. T. Colin Campbell, who is vegan, says that whole foods are best and that whole soyfoods such as edamame are at the top of his list as far as health value is concerned, which is the case for other whole legumes as well.

It is obvious that once any food—soy included—is taken apart and reconstituted into another product, it is surely going to detract from its original value. According to Dr. T. Colin Campbell, there is no question about that; however, there is much debate as to the extent to which it detracts.

> I prefer the whole foods, for sure, but for people who consume the processed soy products, I don't know, how can I say it? I don't really tend to use them, myself, but then again if I'm going to be a serious scientist, my question is to myself, "Where is the evidence that compares these kinds of foods? Reliable evidence?" I can't find it.

CHAPTER 4

Rumor: Heart Health Benefits of Soy in Question

"Cholesterol is a health-promoting substance."

—Chris Masterjohn, Weston A. Price Foundation

Heart health and the prevention of cardiovascular disease (CVD) is the most studied of all the benefits associated with consuming soy.

Heart disease is the leading cause of death in the US and anything that promises to reduce the risk of developing heart disease is sure to attract attention.

In 1999, the FDA approved a health claim for soyfoods. The health claim stated, "Twenty-five grams of soy protein per day, as part of a diet low in saturated fat and cholesterol, may reduce the risk of heart disease." It has been noted that blood cholesterol levels are lowered at intakes less than twenty-five grams a day and that threshold was set, not because there was data showing lesser amounts were ineffective, but because research studies never used less than that amount.

The Weston Price Foundation (WAPF) considers cholesterol to be an essential dietary nutrient and that egg yolks and liver are among the most nutritious of food choices. Their battle cry, "The War on Cholesterol Is a War on Your Freedom," speaks to a counterintuitive endorsement of all manner of animal products, particularly red meat and organ meats, which are internal organs, such as the heart, kidneys, liver, and such.

According to the WAPF, high-fat, cholesterol-laden meat and dairy products are actually good for your heart, while high-fiber, low-fat, antioxidant-rich plant foods such as soy are "poison."

Animal products do not contain a shred of fiber, which is necessary for proper digestion and elimination. Fiber is found only in the cell walls of plant foods, such as legumes, whole grains, fruits, vegetables, nuts, and seeds, with beans at the very top of the fiber-rich food chart. The role of fiber in the diet is related to maintaining the health and integrity of the bowel.

The slower movement of food through the digestive tract is an indication of poor digestion. This often causes food fermentation, which is where gas formation occurs. In time, the fermenting foods decompose and this causes toxic chemicals to form.

A preponderance of research has shown a high-fiber diet may help prevent cancer, heart disease, and a number of other serious ailments.[1 2 3]

Once referred to as "roughage," the American Dietetic Association recommends a daily intake of up to thirty-five grams of fiber per day. The average American consuming the typical American diet, which relies heavily on the consumption of beef, fowl, fish and dairy products, is woefully deficient in dietary fiber.

Americans on the Standard American Diet (SAD) get about 15 grams of dietary fiber a day; that is less than half the minimum daily requirement, which is 31.5 grams of fiber a day.

The Food and Drug Administration (FDA) recognizes the importance of adequate fiber intake and requires the amount of fiber to be listed on the nutrition facts panel of product food labels. Based on scientific evidence, the FDA approved four claims related to fiber intake and a lowered risk of heart disease and cancer in the late 1990s.[45]

One of the FDA's approved health claims singled out soluble fiber: "Diets low in saturated fat and cholesterol and rich in fruits, vegetables and grain products that contain fiber, particularly soluble fiber, may reduce the risk of coronary heart disease."

Legumes are rich in both soluble and insoluble fiber, which are equally important and have different health benefits. Insoluble fiber is basically undigested roughage that moves the food along and decreases transit time in the digestive tract. Wheat bran, nuts, whole wheat flour, and many vegetables including greens, peas, corn, bell peppers, eggplant, celery, onions, and garlic are good sources of insoluble fiber.

Soluble fiber stabilizes blood sugar levels and improves cholesterol levels. Soluble fiber absorbs water to become a gel-like substance in the gut, which actually absorbs cholesterol and that makes a diet high in soluble fiber very beneficial for heart health.

Soluble fiber is found in oats, peas, beans, apples, citrus fruits, carrots, and barley. Soybeans and soyfoods like tempeh, textured soy protein, and soy flour are rich sources of in soluble fiber. Just one cup of soybeans contains 40 percent of the daily value (DV) for fiber.

Dr. John Erdman is a Professor of Food Science and Human Nutrition, Internal Medicine, Division of Nutritional Sciences at the University of Illinois at Urbana. He also holds an Endowed Nutrition Research Chair.

According to Dr. Erdman, soyfoods have a dual beneficial role in nutrition. He says:

> If an individual consumes soy based foods as a protein source in place of animal based foods, which are the source of dietary cholesterol, they already benefit.
>
> I think it could be relatively easily done even for an omnivore who is not a vegetarian. Replacing cholesterol-raising foods with soy means less dietary cholesterol and a better balance of fatty acid profiles as well, assuming you are consuming whole soyfoods.
>
> Consuming soy Isolate or concentrate, which has no fat in it, also has the benefit of lower fat consumption. So, that means that mere substitution is a benefit to overall lipids in the blood.
>
> Secondarily, there are probably hundreds of publications at this point that have looked at feeding soy, usually in comparison to casein- or milk-based proteins on lipids in human subjects. Most of those show an inverse relationship of soy and either total or LDL cholesterol.

It is well known among research scientists that the more elevated the level of total serum cholesterol is, the more dramatic the reduction from consuming soyfoods. Claims that the cholesterol reducing benefit of soy is very small are based on population studies that apply the factor across entire samples and average out the results, which is very misleading.

Meta-analysis is a systematic method of evaluating statistical data that combines the results from separate but similar pre-existing studies to obtain a quantitative estimate of the overall effect of a particular intervention. The pooled data is evaluated for statistical significance.

Dr. Erdman continues:

> More recent meta-analyses indicate effects that seem small, but they are there, in other words a one to three percent reduction in serum cholesterol in a population is still an important factor.
>
> Those meta-analyses are combined. There are numerous different trials that are done by different labs in different locations, under different conditions, using different soy products and, most importantly, are interventions in people with different levels of baseline cholesterol.
>
> I think that many of the studies that are null, that don't show any effect on serum cholesterol, are those that are carried out with people with normal serum lipids. In other words they don't have elevated serum cholesterol. So, one can ask the question whether you would expect a reduction in serum cholesterol if someone has normal serum lipids.

What is considered normal cholesterol? Dr. Erdman indicated that what would be considered normal thirty years ago would not meet today's standards. "Total cholesterol under 200 milligrams per deciliter (and for LDL cholesterol, under 130 milligrams per deciliter) is now the modern standard for normal cholesterol levels. At one time 250 milligrams per deciliter or even 280 milligrams was thought to be 'normal'; over the years it keeps coming down."

Apparently the same holds true for blood pressure. Dr. Erdman: "The cutoffs come down as we learn more about the fact that in a very large population there is a risk of being between 200 and 230, or between 230 and 250, you do get increased risk."

Heart disease is the number one killer of women and men in the United States. Annually, more than a million Americans have heart attacks and about a half million people die from heart disease. High blood cholesterol is a major risk factor for heart disease, which means that the higher the blood cholesterol level, the greater the risk for developing heart disease. Doctors recommend a total cholesterol level below 200 milligrams/dL.

> I think it is a fact of the metabolism of those individuals. It's much easier to reduce a component like serum cholesterol that's extremely high than if it's near normal, because there are homeostatic mechanisms that work to try to keep a number of blood parameters within a small narrow range.

In one of Dr. Erdman's earlier trials, an intervention study conducted at the Danville Veterans Administration Hospital, one of the individuals screened for the study had serum cholesterol level that was in the 600 milligrams per deciliter range. Obviously, this man was in deep trouble and Dr. Erdman had serious issues with admitting this individual into the study. He was told that they could not accept him into the trial and that he really needed to see a doctor. This was in the late 1980s.

The gentleman was extremely distraught and eventually they did admit him into the trial. While the team could not include him in the final sample, there is a notation in the manuscript that this man's cholesterol came down to around 200 milligrams per deciliter. Consuming soy really worked, and to the degree where this patient's cholesterol plummeted.

According to Dr. James Anderson, chief investigator on the landmark 1995 meta-analysis on soy and heart health,[6] soyfoods are some of the healthiest foods you can put on the table. This is because they help fight what Dr. Anderson calls the big five: heart disease, stroke, diabetes, obesity, and high blood pressure.

Soyfoods are rich in fiber and Dr. Anderson believes that soy is a healthy, low-fat, cholesterol-free alternative to animal products.

"I've looked carefully at cholesterol. I am sure soy protein lowers blood cholesterol. The data is very clear to me. I've looked at blood pressure doing a meta-analysis on soy protein lowering blood pressure and it is the peptides that are doing that."

In addition to soy protein there may be other components in soy that can benefit cardiovascular health. Dr. Erdman says:

> There may be some other benefits for heart health that would be attributable to the isoflavones in soy.
>
> There are a few papers coming out here and there regarding blood pressure reduction and while this probably needs further study, there are some associations of the isoflavones and other polyphenols with reduction of blood pressure, which is the other accepted marker of cardiovascular disease.

CHAPTER 5

Rumor: Soy Contains Dangerous Substances

Anti-soy claim:

***Soybeans contain large quantities of potent "anti-nutrients" including enzyme inhibitors that block the action of trypsin and other enzymes and phytic acid, which can block the absorption of essential minerals.*[1]**

According to anti-soy activists, soybeans contain compounds that are dangerous for human and animal consumption. They claim that soy is full of anti-nutrients such as enzyme inhibitors that block the action of trypsin and other enzymes essential to the digestion of protein.

The kidneys filter waste from the blood and one symptom of kidney disease is an excess of protein in the urine. Dr. Roberta Gray is a pediatrician with a subspecialty in Nephrology, a branch of internal medicine and pediatrics dealing with function and diseases of the kidney. Dr. Gray explains:

> I've had two different types of practice settings: General academic pediatrics and the other in my subspecialty practice; my subspecialty is kidney disease. I have looked at soy products in the general context as a sub-specialist and really found the soy beverage and protein substitute products to be extremely useful in my nephrology/kidney disease practice.

Phosphorous is an important issue for people with any degree of abnormal kidney function. An individual needs to have excellent kidney function in order to get rid of phosphorous. If the body cannot get rid of phosphorus, it will team up with calcium, which lowers the blood calcium level, causing another form of bone disease. This is called chronic kidney ricketts or metabolic bone disease (CKD-MBD).

Metabolic bone disease is a common complication of chronic kidney disease. CKD-MBD occurs when the kidneys fail to maintain proper levels of calcium and phosphorus in the blood, leading to abnormal bone hormone levels.

As a pediatric nephrologist, Dr. Gray sees many children with chronic kidney conditions. She found that children with chronic kidney failure who are not yet on dialysis could see great improvements with protein restrictions, so Dr. Gray began recommending a diet based on plant protein instead of animal protein. Dr. Gray says:

> Phosphorous is poison to a kidney failure patient and meat and dairy products are very high in phosphorus. So having soy as an option both for milk substitutes and meat substitutes is extremely beneficial. I have prescribed a vegetarian diet, including soyfoods, meat alternatives, and plant-based foods for many, many years and for my patients who had an identifiable kidney problem that was likely to progress.

The fact is that chronic kidney disease (CKD) is increasing.[2] Individuals with CKD are unable to eliminate excess phosphorus, which can build to toxic levels in the body. A study conducted at the Cleveland Clinic demonstrated that patients eating a vegetarian diet absorbed less of the mineral nutrient than those eating meat.[3]

Study participants were divided into groups consuming either plant-based or meat-based meals that contained equal amounts of protein and phosphorus concentrations. The patients eating a diet of vegetables only had lower blood phosphorus levels and decreased phosphorus excretion in their urine after one week.

Cow milk contains twice the amount of phosphorus as soymilk and in the Western diet an abundance of phosphorus is consumed from meat protein and dairy products.

The researchers wrote, "Although we advise our CKD patients to follow a phosphate-restricted diet, we rarely discuss the protein source of phosphate, which we have now demonstrated to be important."

The researchers noted that grain-based diets have lower phosphate-to-protein ratios and that much of the phosphate in vegetables comes from phytate—a nutrient not absorbed in humans.

Dr. Gray explains:

> I can't say that all nephrologists recommend that the protein be vegetable protein. I recommend it. And the fact of the matter is that the parents of my patients have told me that their children who have to limit their protein are so much more satisfied when they eat a plant-based diet.
>
> That is because of the fact that they have so much more bulk – the fiber—and other things in their diet that they felt so much more satisfied. Parents would tell me that their children weren't always screaming, "I want a piece of baloney."

In Dr. Gray's experience, those patients with a compromised renal system who would have difficulty filtering toxins from the bloodstream can more easily digest soyfoods.

"When these children have been on a meat-based diet and because of a kidney problem went to a plant based diet; they ended up being infinitely more satisfied and were meeting all of their nutritional requirements."

Another benefit of prescribing the soy diet for her juvenile patients is that parents are so impressed that the whole family adopts these healthier habits. Dr. Gray continues. "Many children develop kidney stones due to their meat-heavy diets. Moving them toward a plant-based diet is an important part of management of patients with this condition."

If the soy antagonists had the facts correct, it would follow that soy products would be damaging to the kidneys. However, Mayo Clinic urologist, Erik Castle, MD recommends soy protein products, such as tofu, tempeh, soymilk, and soy yogurt in place of animal products for kidney patients.[4]

What Are Anti-Nutrients and Why Are They There?

Anti-nutrients are naturally occurring compounds that are common at some level in a number of plants for various reasons. It is during seed germination that certain enzymes become active, which play a role in stimulating the development of the seed and stimulating early plant growth.

Phytic acid is one such compound found in many cereals and legumes that allow the plants to store phosphorus. These anti-nutritional compounds such as amylase inhibitors, lectins and trypsin inhibitors are present in legume seeds and also function to protect plants against predators.

Raw soybeans contain trypsin inhibitors and phytic acid. These are compounds that are found in almost all plant foods and conventional cooking will deactivate the anti-nutrient activity of these substances.[5]

Studies have shown that soaking leads to a significant reduction of phytate, which is found in millet, maize, rice, and soybeans and is very effective in combination with other processes, such as further cooking.[6]

Trypsin inhibitors, sometimes called protease inhibitors, are proteins that diminish the availability of trypsin, which is an essential enzyme for humans and other animals.

Phytic acid is found in grains, legumes, nuts, and seeds and forms insoluble complexes in the presence of minerals such as calcium, zinc, iron, magnesium, and copper and inhibits their absorption.

Though present in the bran of many grains and in the hull of all seeds, the inhibiting effect of phytic acid is deactivated by soaking or fermenting. All beans, including soybeans, must always be first soaked and then subject to lengthy cooking; it is safe to say that raw grains and beans would not be in any way palatable otherwise.

As is the case with trypsin inhibitors and other anti-nutrients, proper cooking dramatically reduces the levels of phytic acid. The correct method of food preparation regarding beans and grains, including removing the hull and soaking the beans, has been common knowledge for more than ten thousand years.

Flavonoids, a subclass of polyphenols, are among the most well known of the polyphenols and are widely distributed in nature.[7] Polyphenols are a class of antioxidants thought to have a substantial and well-documented health benefits.

Genestein, an Isoflavone found in soy, is a member of the flavonoid family of compounds. While flavonoids have been shown to have a positive impact on health, they can also interfere with nutrient absorption.

Flavonoid compounds include tannin, which is present in tea, wine, and fruit and is well known to reduce the absorption of iron and zinc and may also inhibit digestive enzymes and affect proteins.

However, polyphenols such as tannins also have anticancer properties and beverages such as green tea, which contains large quantities of these compounds, are considered very beneficial to our health despite their anti-nutrient properties.[8]

Many plant foods, including soybeans, contain certain anti-nutrient compounds when raw; however, proper preparation and cooking will nullify the effects.

In fact, about 90 percent of the trypsin activity in soybeans is destroyed by cooking and further reduced by presoaking and removal of the exterior hull. In the case of miso, tempeh, soy sauce, and other fermented soyfoods, fermentation also has a deactivating effect.

Traditional cultures have been soaking and fermenting legumes and grains for many centuries. The first step in the preparation of soybeans, whether in the home kitchen or for commercial processing, is to soak the beans for up to eighteen hours. The length of soaking depends on what you are doing with them. When making soy nuts, the beans are soaked for at least three hours, whereas for soymilk or tofu, the beans are soaked for at least eight and up to eighteen hours.

The presoaked soybeans are then cooked for a sustained period of time—at least three hours if you are making a dish with whole or mashed soybeans. Needless to say, unprocessed (neither cooked or soaked) legumes and grains are not consumed in their raw state. This fact effectively renders the anti-nutrient argument null.

Anti-soy claims:
Fermented soyfoods are better for you.
Unfermented soyfoods are not safe to eat.
Traditional Asian cuisine uses only fermented soy products.

There are some nutritional differences between fermented and non-fermented soyfoods. Many traditional Asian soyfoods such as miso, natto, and tempeh are fermented; however, tofu and soymilk are not. While fermented soyfoods are believed to be less allergenic,[9] fermenting neither increases nor decreases the bioavailability of the isoflavones.[10 11]

Fermentation does reduce the phytate content of soyfoods and will, to some degree, improve the absorption of minerals, although the extent is not yet clear.[12 13]

Researchers in Japan evaluated more than thirty thousand men and women, over a period of fifteen years and found that people who consumed the most soyfoods had a significantly decreased risk of stomach cancer, the leading cause of cancer-related deaths. Specifically, higher intakes of non-fermented soyfoods such as tofu, soymilk, and edamame were significantly associated with a lower risk of stomach cancer and were found to actually have a protective effect against developing the cancer.[14]

Their findings suggest that a higher intake of non-fermented soyfoods is associated with a lower risk of stomach cancer in both men and women. The study found that there was no significant association between the intake of fermented soyfoods and a risk of stomach cancer. These results suggest that a high intake of soy isoflavone, mainly non-fermented soyfoods, have a protective effect against stomach cancer.

According to Bill Shurtleff, a foremost authority on soybeans and soyfoods, Kaala Daniel and Sally Fallon both have a vision of a world where people eat more meat and raise it at home in their back yards.

"They are very down on the whole isolates, concentrates, and so forth . . . which I tend to be in general. However, I am not opposed to these products if they can help people get off meat . . . then I am not opposed to it."

Shurtleff continues:

> Whereas they would be categorically opposed to these more processed soy products because they are more non-traditional foods, they've also developed, for the first time ever, the concept that fermented soyfoods are better than non-fermented soy. There is no evidence to support that at all.
>
> Fermented soyfoods are great! You can make a list of fifteen different advantages that come from fermentation, but that doesn't mean that there's something wrong with edamame. That doesn't mean that there's something wrong with tofu.

Anti-soy claim:
Soybeans contain haemagglutinin, a clot promoting substance that causes red blood cells to clump together.[15]

The issue raised here regards a type of protein called lectin, which is also known as haemagglutinin. All plant and animal products contain quantities of lectin in varying degrees. Grains, dairy products, tree nuts, and nightshade vegetables such as tomatoes, potatoes, eggplant, peppers, and legumes all contain lectin.

The lectins in plant seeds are thought to play a role in the process of seed germination, the earliest stages of growth where a tiny plant emerges from the seed and begins to grow. The concentration of lectins in plant seeds, which may contribute to the survival of the seed, decreases over time as the plant grows and matures.

Noting that humans do not consume raw soybeans, I asked Bill Shurtleff what he would say about the claim that soy contains dangerous toxins.

> They fail to make that distinction again and again and again when they are talking about toxins in soy. They don't distinguish between those that have been cooked and those that are uncooked. And to the person who doesn't know the subject, it's easy to slip it by them when you talk that way.
>
> The seed does not exist for human beings, it exists for its own survival and these toxins are there to insure the survival of the seed and the plant. They are substances that are there to protect the seed from insects and such. There's hardly a food that you can find that doesn't have its own armory of one sort or another of toxins.

One of the allegations made by Fallon and Enig of the WAPF is that soy fed to weanling rats caused them to fail to grow normally. The article appears on the WAPF website with the heading "Soy Alert" and was also published in *Nexus Magazine* in 2000.[16] While there is no reference to a particular study in the article, the authors proclaim that the result of the phantom study was due to haemagglutinin and trypsin inhibitors.

Were plant foods containing lectin or any of these anti-nutrient compounds consumed raw, they would be toxic to both animals and humans.

It really cannot be stressed enough that these claims raised and clouded by the anti-soy rhetoric, are a non-issues because people do not consume soy or any of these foodstuffs in their raw state and the anti-nutrients, including haemagglutinins that occur in raw plant foods, are temperature-sensitive and deactivated by conventional cooking.

According to the FDA, red kidney beans contain ten times the level of lectins found in soybeans; however, these lectins are rendered harmless when cooked properly.

According to the Food and Agriculture Organization of the United Nations:

> The lectins, formerly known as hemagglutinins, are proteins, which possess the ability to agglutinate red blood cells (firmly stuck together to form a mass). They are widely distributed in plants and some, such as the castor bean lectin or ricin, are highly toxic. The lectin found in raw soybeans has, apparently, no observable dietary effect, good or bad. Furthermore, it too is easily inactivated by heat.[17]

The anti-soy contingent has gone to great lengths to promote the notion that it is not safe to consume soy unless it has been fermented. The WAPF authors and their allies contend that only fermenting will render these substances inactive.

While fermenting will deactivate these compounds, these anti-nutrients are heat sensitive and fermentation is not the only effective method of preparation.

According to Dr. Mark Messina, the most prominent of soy researchers, the calcium in soy is actually well absorbed, in spite of the phytates:

> Critics of soyfoods say that soy is high in phytates, which inhibit absorption of iron, zinc, and calcium. But the absorption of calcium from soyfoods is actually surprisingly good given the phytate content of those foods.[18 19]
>
> Not only that, but a number of studies have shown that the isoflavones in soyfoods protect bone health[20] and that soy protein when substituted for an-

imal protein decreases urinary calcium excretion.[21] So getting calcium from soyfoods that are either naturally rich in this nutrient or are fortified with it, seems like a very good idea.

That means that soyfoods can contribute to bone health and consuming tofu that has been coagulated with calcium sulfate and soymilk, which has been fortified with calcium, will supply additional sources of bioavailable calcium for the soyfoods aficionado.

Dr. Messina continues:

> But it is true that, all other things being equal, phytates inhibit the absorption of iron and zinc. Soybeans are rich in phytate and vegan diets are especially high in phytate. It is very well documented that vegetarians absorb iron less well than meat eaters and have lower stores of iron in their bodies. But the implications of this aren't clear.
>
> Vegetarians don't appear to be any more likely to actually develop iron deficiency.[22] And, because high levels of iron may raise risk for heart disease, it may be that having lower but adequate stores as vegetarians do is the ideal situation.[23]

On the question of whether it is safe to consume soy, noted nutrition researcher, Dr. T. Colin Campbell had this to say:

> I would consider soy to be a healthy component of a plant-based diet, but not to the exclusion of all of the other beans and peas and things like that, obviously. Soy is a complement to that group of foods and it is useful.
>
> I hear a lot of shouting and yelling going on in the popular press about things like this, but so much of that is coming from the industry that you don't

> know what to believe. I don't pay a lot of attention to that stuff, because I see claims made on both sides and I want to know what the science is and regarding claims that consuming soy is dangerous? The science is not there.

What about the role of soy in replacing unhealthy food choices in the diet? Dr. T. Campbell says:

> Absolutely! The science of nutrition is grossly misunderstood by professionals and the public, alike. It is time that we begin to recognize science for what it is and understand it properly. There is no question in my mind that the argument is over as far as what kind of foods we should be consuming. It has to be whole, plant-based foods, and that's it.

CHAPTER 6

Rumor: Eating Soy Causes Thyroid Problems

"Soyfoods do not seem to cause hypothyroidism. They do not seem to affect the health of the thyroid at all."

—Neal Barnard, MD, Physicians Committee for Responsible Medicine Washington, DC

The function of the thyroid gland is to convert the iodine found in food into thyroid hormones. These hormones, thyroxine (T4) and triiodothyronine (T3), enter the bloodstream, which then transports them throughout the body. Thyroid hormones play a critical metabolic role in the conversion of oxygen and calories into energy. Every cell in the body depends on these thyroid hormones for the regulation of metabolism.

Iodine is a chemical element that is plentiful in seawater and in various minerals found in the soil. Sea vegetables are an excellent source of iodine and strawberries are also a good source; however, most foods contain low levels that vary according to environmental factors, such as the concentration of iodine in the soil. Some of the richest sources of iodine are products that include iodized salt and bread made with iodate dough conditioners.

Goitrogens are naturally occurring substances found in many foods. These compounds can suppress the thyroid gland by interfering with the uptake of iodine, thus making it more difficult to produce these hormones. Goitrogens are found primarily in cruciferous vegetables, and legumes. Cauliflower, cabbage, broccoli, spinach, turnips, and kale contain goitrogens, as do lima beans, sweet potatoes, peaches, strawberries, millet, and soybeans, among others.

Goiterogenic foods can impede the absorption of iodine and thus slow down thyroid function, particularly when consumed raw. In extreme instances these compounds can cause the thyroid gland to enlarge, a condition called goiter. That being said there is no evidence that consuming soy in any way causes thyroid problems in healthy individuals with sufficient levels of iodine in their diet. There is a concern regarding goiterogenic foods where an iodine deficiency or thyroid dysfunction is already present.

There may be an issue for some individuals regarding goitrogens and thyroid function and anyone with thyroid disease may wish to adjust or limit their intake of goitrogenic foods. It has been suggested that individuals with a thyroid hormone deficiency may tolerate an eight-ounce serving of cruciferous vegetables and four-ounce serving of soy two or three times a week.

According to nutrition researcher Dr. Neal Barnard:

> Soyfoods do not seem to cause hypothyroidism. They do not seem to affect the health of the thyroid at all. And yet, at the same time it may well be true that isoflavones or many, many, many foods do affect the absorption of oral preparations such as thyroid hormone. And so, you can take them at a different time of day.

He continues:

> But the other thing that should be said is that doctors never just put a person on thyroid hormone and leave them alone. They track their hormone levels over time. And so, if your hormones are a little bit low, they modify the dose. And so the fact that one or another food might increase or decrease the absorption of thyroid hormone a little bit doesn't have any clinical significance.
>
> These foods have very strong nutritional value and are known to prevent many health problems and if you do have a thyroid issue, it is worth asking your medical professional about including them in meal planning on a limited basis.

Soy consumption and its effect on the functioning of the thyroid gland are issues that have been studied for almost three-quarters of a century.[1] Researchers reported that when raw soybeans were added to an iodine-deficient diet of rats, they developed goiter.[2] Animal experiments continued and further study revealed that the addition of iodine to the diet could prevent goiter.[3] This became an active area of research, which led to the belief that soybeans contain a goitrogenic factor.[4]

In the 1950s there were about a dozen cases of goiter reported among infants who were fed infant soy formula, which was derived from soy flour. The soy infant formula was reformulated with soy protein isolate and fortified with iodine. The original soy infant formula has not been in use since that time and there have been no infant formula related cases of goiter reported in studies to date (see soy infant formula discussion in Chapter 7).

Articles that claim soyfoods cause thyroid disease often cite animal studies with rats fed gigantic amounts of genistein, an isolated extract of soy. Nutrition scientists conduct research such as this in order to help them determine the role of extracts of various components in certain foods. The problem with the anti-soy rhetoric is in the all-encompassing indictment of soyfoods as a cause of thyroid disease.

Soy or other foods deemed to be goitrogenic do not cause thyroid disease in healthy individuals with a functioning thyroid. There is no need to be concerned about consuming spinach, cauliflower, Brussels sprouts, leafy greens, or soy. While the goitrogenic properties of these and other wholesome foods should be a consideration for those with thyroid dysfunction, the best dietary strategy is to include adequate iodine.

Soy protein isolates are used throughout the food industry for nutrition and functionality in power bars, protein drinks, protein powders, and pill supplements. Soy protein isolates are also found in meat analogues, breads and baked goods, breakfast cereals, and many fitness products designed to increase weight or muscle mass.

The relationship between soy, soy isoflavones, and thyroid function in adults has been extensively analyzed and reviewed by nutrition researchers. In the case of isolated soy protein, researchers conducting human trials among postmenopausal women with healthy thyroids showed no effects of isoflavone supplements on thyroid function.[5]

Postmenopausal women tend to have a higher incidence of hypothyroidism than the general population, and the results of this 2003 study suggest that soy isoflavones do not adversely affect thyroid function.

Soybeans, like all legumes, whether packaged for consumers or commercial processors, have explicit preparation instructions and are not intended for raw consumption.

Goitrogens appear to be heat-sensitive and cooking is thought to lessen their availability by as much as a third. In the manufacture of soyfoods, soybeans undergo a lengthy presoak and cooking in advance of further processing. With the exception of lentils, legumes must be presoaked before cooking and would be inedible if raw.

Medical and nutrition experts agree that in the absence of a thyroid disorder, consuming soy does not cause thyroid problems. Healthy individuals who are not deficient in iodine can enjoy these healthful and nourishing foods without concern for thyroid health.

Iodized salt is a common dietary source of this mineral and most multivitamin and mineral supplements provide the recommended daily allowance for iodine. There is no need to avoid cruciferous vegetables or soyfoods because of a concern for thyroid health. The preference is to consume a balanced diet, which includes an adequate intake of iodine.

According to Dr. Mark Messina:

> There is no reason to restrict soy consumption over concerns about the impact on thyroid function, but of course all people, regardless of their dietary pattern, need to consume sufficient amounts of iodine.
>
> Any concerns about the effect of soy on thyroid levels can be definitively addressed by having thyroid hormone levels measured. Even this step is not unordinary, since the American Thyroid Association recommends that all people have their thyroid

hormone levels checked every five years beginning at the age of thirty-five.[6]

CHAPTER 7

Rumor: Soy Formula Is Dangerous for Infants

"An infant exclusively fed soy formula receives the estrogenic equivalent (based on body weight) of at least five birth control pills per day."

—Sally Fallon, Weston A. Price Foundation

One of the most disturbing claims emanating from this controversy has been the charge that feeding with soy infant formula is very dangerous for babies.

Parents were horrified to hear that they may be feeding their infants a diet that could be detrimental to their child's health and, as a result, many stopped feeding with soy formula. The claims came out of New Zealand and, as is often the case, were based on theories regarding the phytoestrogens in soy.

These anti-soy claims alleged that soy infant formula would adversely affect the thyroid, causing goiter, and would create havoc in the reproductive system. The theory asserted that feeding soy formula to infants has long-term effects, causing of all manner of sex organ and hormonal abnormalities including testicular atrophy and a tendency toward homosexuality.

The anti-soy contingent frequently cites cases of goiter diagnosed in about a dozen infants fed infant formula made from soy flour in the 1950s.[1] By the early 1960s the soy infant formula base was modified and fortified with iodine. There has not been a case of goiter attributed to soy infant formula since that time.

Soy infant formula, like all food products, is regulated by the FDA and has been shown to promote normal growth and development as well as cow milk formula. Millions of infants worldwide fed soy infant formula are now adults and there has been no pattern of abnormalities yet to emerge.

An infant subsists on a diet that consists exclusively of milk. Parents of the child who is not breastfed must choose between cow and soymilk formula, neither of which was intended by nature.

Dr. Mark Messina points out that soy formula produces normal growth and development. "Since the early 1960s, about twenty million infants have used soy formula for various lengths of time."

It is important to note that there have been no reports of hormonal abnormalities among individuals who were fed soy formula as infants. That would include the many millions of individuals who have been fed infant soy formula since the mid-twentieth century.

A major study published in the *Journal of the American Medical Association* in 2001 found that infants fed soy formula grow to be just as healthy as those nourished with cow milk formula.[2]

Soy protein has been used for infant feeding in one form or another in America for almost a hundred years. While soy nutrition in infancy was common for centuries in Asia, the earliest use of soy as an infant formula in the US was in 1909.[3]

There is no question that breastfeeding is the best possible regimen for babies, particularly in the first year of life. It is also true that researchers have known since the early 1990s that exposure to cow milk in infancy can increase the risk for type 1 diabetes.

Type 1 diabetes occurs when the body's immune system attacks and destroys insulin producing cells in the pancreas and usually appears sometime in childhood or adolescence.

Studies demonstrate that early exposure to cow milk in infancy can increase the risk for type 1 diabetes. In 1992 the *New England Journal of Medicine* published a report that implicated cow milk protein as a possible trigger of this autoimmune response.[4]

While we don't hear a lot about the JAMA paper from 1992, the disturbing allegations regarding soy infant formula continue to be an issue of debate and controversy.

Thomas Badger, PhD, Neuroendocrinologist, and Director of the Arkansas Children's Nutrition Center (ACNC), is conducting the Beginnings Study, a long-term clinical study of the development, nutritional status, and health of infants from birth to six years of age and through puberty.

His subjects are divided into three groups: breastfed infants, children fed cow milk formula, and children fed soy formula. Preliminary data of Dr. Badger's study show that children nourished by breastfeeding or either of the two major formulas, cow milk formula or soy formula, grow and develop in essentially the same way.

This study is the first to prospectively determine the differential characteristics of growth, body composition, and brain development of cow milk formula fed, soy formula fed, and breastfed infants.

Numerous studies have been conducted comparing the effects of soy formula on growth and measuring weight, length, and head circumference with cow milk fed and breastfed infants; however, there have not been any prospective, longitudinal studies over such a long developmental period as the Beginnings Study. Previous studies also did not have a specific emphasis on brain development and function, i.e., behavior, cognition, psychomotor and language development, and body composition.

Although the study will not be completed until all the children have gone through puberty, estimated to be around 2025, they have begun to publish on data sets through the first years of life.[5][6]

Dr. Badger explains:

> In general, our preliminary findings thus far have shown that growth and development of soy formula–fed and cow's milk formula–fed infants do not differ significantly; however, there are significant diet group-related effects on body composition, such that the profiles of fat mass, fat free mass, and bone mineral content (BMC) differ significantly over the first twelve months of life.[7]
>
> The purpose of our study is to determine if we can find any differences between these three modes of feeding children, and the supposition would be that if there would be any of these adverse effects—all these crazy reports that are being spread over the Internet—then we should be able to pick them up with careful study like in the one we are doing.
>
> In our study with lots of kids we are looking at lots of different systems, from brain development to hormonal development, and so far we haven't found anything at all . . . and really to be honest with you, I don't expect that we are going to find anything.

The information implicating soy infant formula feeding as the cause of numerous abnormalities is based on animal studies (in vivo). The in vivo model introduces massive amounts of an isolated component to a specially bred animal, usually a rodent such as a mouse or a rat.

The subject is then observed and the results noted and analyzed. Studies such as these are often done to gauge toxicity, generally for dosage information and pharmacological applications, and do not relate to the normal dietary human experience.

If we were to purify out various components of any food and inject or consume those substances in massive quantities, the results would certainly be toxic. Components extracted in this way do not appear in nature and anything in the human diet taken to extremes will cause illness and even death in some cases.

For instance, when consumed in appropriate amounts, something like salt is a harmless and very common dietary ingredient. However, if you were to drink a mixture of salt and water in a strong concentration, you would soon begin to feel very sick.

In fact, according to Dr. Badger, you could well die from drinking too much concentrated salt water and that really is not an appropriate model to study the physiological effects or the normal development of children. That is a model designed to study toxicity.

Dr. Badger continues:

> Some of these studies where a mouse is injected with isoflavones and they develop birth defects or they develop some kind of condition that would lead to cancer or reproductive failure are so far off the chart.
>
> They are really not modeling what occurs in people, because, for example, when these studies are compared to soy infant formula those kinds of conditions do not exist in children.

There has been a fair amount of hysteria surrounding the claims that the isoflavones in soy formula exert an estrogenic effect on infants that causes abnormalities in the reproductive organs. Scientists at the Arkansas Nutrition Center at the University of Arkansas for Medical Sciences published a study in February 2010 in the *Journal of Pediatrics* investigating whether feeding soy formula was creating an adverse effect on the reproductive organs of little boys and little girls.[8]

The data in this study, which measured reproductive organ size by ultrasound in infants at age four months, did not support major diet-related differences in the soy formula–fed infants. The researchers concluded that there was no evidence that feeding soy formula exerts any estrogenic effects on reproductive organs.

Interestingly, these scientists did find some evidence that ovarian development may be advanced in cow milk–fed infants and that testicular development may be slower in both cow milk–fed and soy formula–fed infants as compared with infants who were breastfed.

The American Academy of Pediatrics updated their position paper from 1998 on soy infant formula stating that while soy formula may have no advantages over cow milk formula, for term infants "there is no conclusive evidence from animal, adult human, or infant populations that dietary soy isoflavones may adversely affect human development, reproduction, or endocrine function."[9]

Are soy formulas safe? Dr. Badger answers:

> Well, that's what we are testing right now. That's what we are studying. We are not intervening in any way. And so far we have found absolutely no significant differences. There are some slight differences. For example, little boys fed milk formula have a tendency to be grossly overweight at a very early age. I am not saying all of them, but a significant number of them.
>
> In the soy infant formula group, we find that these kids tend to be leaner, so they're all within the normal range. We don't find any of these really over-

> weight infants in the soy formula group. There are a higher percentage of children fed cow's milk formula who are heavier.
>
> For instance, in our current study there are about 22 percent of cow's milk formula–fed boys that are over the 95th percentile. That number should only be 5 percent and that means that there are about four times more children who are grossly overweight than you would expect.
>
> Now is that good or bad? Until we finish the study and take these kids all the way out and study their body composition and ask questions about how they come out at age six years, we're not really going to know. But so far, that looks like an advantage to soy formula, not an adverse effect at all. So it's that type of thing that we are actually finding.

This is a big issue. . . . A lot of people have said a lot of things about this but have never actually studied it, and we're actually studying it. And so we think that before we go off and get everybody fired up, let's get the facts. And once we get the facts on children that have actually been fed soy formula and not just animals that have been injected with or given these purified compounds, let's see what the real facts are.

Haley Curtis Stevens, PhD, is the Scientific Affairs Specialist for the International Formula Council (IFC), an association of manufacturers and marketers of formulated nutritional products such as infant formulas and adult nutritional products. I asked Dr. Stevens about the recent campaign to discredit soy infant formula.

> Soy infant formulas are a safe and effective alternatives for infants whose nutritional needs are not met by human milk or formulas based on cow's milk. Modern soy formula, which is based on soy protein isolates,

> has been used for over fifty years without reports of negative reproductive or developmental effects.
>
> With a long history of providing infants safe, effective and trusted nutrition, soy formulas are a recognized, acceptable alternative to milk-based formulas.
>
> Recent theoretical concerns over the safety of isoflavones contained in soy formula are based largely on contrived and artificial animal models that are now known to be inaccurate representations of human metabolism of plant phytoestrogens.

Dr. Stevens emphasized that there has been no valid clinical evidencc that isoflavones in soy formula pose a significant health risk to humans. It is important to note that soy infant formulas meet a critical need in that they provide options for infants with clinical indications that preclude feeding with cow milk–based formula. Soy infant formula also provides a safe alternative for parents whose cultural or religious practices necessitate a cow milk–free formula.

Roberta Gray, MD, is pediatrician and clinical professor who taught at institutions such as Duke University and is now doing physician and patient educational consultations in the southern US. Dr. Gray explains:

> From the standpoint of a pediatrician, for the entirety of my career in pediatrics the use of soy formula for a variety of reasons for infants has been extremely widespread and regarded as very safe, very beneficial, and very healthful by pediatricians for generations.
>
> I know there are some who are concerned about sensitization to soy, like what occurs with cow's milk protein. More recently, there have been concerns about the estrogen-like effects that some soy products may have. However, I am a proponent of soy formula. I've never had any concern about it whatever.

My subspecialty is in nephrology and there are enormous benefits for older children in consuming soymilk and various other soymilk preparations. The type of protein soy contains, the amount of phosphorous and other things are really beneficial to children who didn't have normal kidney function, and actually to a subset of children who have recurring kidney stones.

CHAPTER 8

Rumor: Soy Has a Feminizing Effect on Males

Anti-soy claim:

Consuming soy adversely affects fertility, lowers sperm count, and causes gynecomastia, the swelling of breast tissue in young boys.

The effect of soy isoflavones on hormone levels in the body has been the subject of much speculation. It has been implied that isoflavones in soy have a negative effect on male hormones and may adversely affect male fertility, possibly lessening semen quality and testosterone levels. The theoretical claims that eating soy could cause infertility would seem counterintuitive when considering the five thousand year history of soy in Asian culture.

China is still the only country on earth that found it necessary to impose mandatory birth control on its population. China, the birthplace of tofu, where the soybean is venerated and consumed early and often, was forced to take steps to control its population. In 1980 the government instituted a one-child limit on all couples of childbearing age. So it is safe to say that procreation, sperm count, and fertility are not an issue in Asia.

While these compounds have been described as plant estrogens, they are not the same as human estrogen, and studies have shown that they do not adversely affect testosterone levels in men.[1] Jill Hamilton-Reeves, PhD, RD, was the principal investigator on a meta-analysis of nine clinical studies of theoretical concerns that isoflavone exposure from either supplements or soyfoods could have feminizing effects on men.

The results, published in the journal *Fertility and Sterility*, were contrary to previous animal studies and found that isoflavone supplements and isoflavone-rich soyfoods did not exert an effect on testosterone levels. Dr. Hamilton-Reeves explains:

> We found that there was no significant effect of soy protein or soy isoflavones on testosterone or some of

the related hormones like free testosterone, which is calculated or free androgen index, which is another calculated measure or sex hormone–binding globulin. The main findings were around testosterone.

The reason we did the meta-analysis was to systematically and mathematically, qualitatively look at the data and see if you look at it all together, where does it lie?

And the two reports that suggested that the testosterone had changed did not have baseline measurements. And so, for study design, we don't consider that as strong of a design, because you have to consider each person's hormone metabolism in order to compare them to themselves.

Animal studies have been taking a lot of fire from concerned activists and some researchers who contend that, aside from the inherent cruelty, studying other species does not provide answers that relate to the human experience. Oftentimes, rodents or primates are injected with massive levels of isolated compounds that would not appear on anyone's dinner table. Dr. Hamilton-Reeves states:

Metabolism is different between species and if they are injecting then they are not accounting for what happens in the gut. The gut metabolism is critical to the efficacy of isoflavones for therapeutic purposes. Again, it's getting context and appreciating the complex physiology of the animal.

Dr. Hamilton-Reeves conducted a clinical study with men who were at high risk for prostate cancer, evaluating testosterone both before and after intervention with a soy protein drink.[2] This study was actually among those reviewed within the meta-analysis published in 2010.

The researchers in this previous study looked at individuals, not population averages, and did not find any changes in testosterone levels. Dr. Hamilton-Reeves says, "This study was on the longer side of intervention studies and our clinician also evaluated individuals and their responses and everyone remained within normal parameters for testosterone."

Jorge Chavarro, MD, is an Assistant Professor of Medicine at Harvard Medical School whose work is investigating the role of diet on human fertility. A continuation of a previous study[3] by Dr. Chavarro and team at Harvard University found male soyfood intake to be unrelated to clinical outcomes. In the follow-up study, couples were recruited between February 2007 and May 2014 and followed to document treatment outcomes including fertilization, implantation, clinical pregnancy, and live birth.[4]

Gynecomastia is a swelling of the breast tissue in boys or men, caused by an imbalance of the hormones estrogen and testosterone. The soy antagonists strongly suggest that consuming soy can cause this rare condition that is mentioned in medical literature.

In May of 2009, *Men's Health* magazine published an article that stated in the opening paragraph that soy was a sinister though perfect protein source that had "the power to undermine everything it means to be male."

The article recounted the story of a man who actually drank three quarts of soymilk a day. This was the only case cited in the article where an individual developed gynecomastia.[5]

Dr. Hamilton-Reeves explains:

> In the scientific press, there really is not a lot of strong evidence that that would happen in the population. There is a case report of gynecomastia and in that report the man consumed about 3 quarts of soymilk a day, which as a dietician I must say is an excessive amount.

> I am surprised that an individual could drink three quarts of any beverage in a day, let alone one as filling as soymilk.
>
> Clearly, excessive intake of any food might produce some sort of effect and with an individual report you do not have statistical power so we cannot be certain whether it was the phytoestrogens or not. This was an individual case, and most of the concerns as to what the media might be pulling from there is mixed data.

A recent study by Dr. Mark Messina, published in the journal *Fertility and Sterility* evaluated the clinical evidence regarding concerns that isoflavone exposure in either supplements or soyfoods has feminizing effects on men.[6]

The study showed that isoflavones do not exert feminizing effects on men at intake levels equal to and even considerably higher than are typical for Asian males. Dr. Messina stated, "In contrast to the results of some rodent studies, findings from a recently published meta-analysis and subsequently published studies show that neither isoflavone supplements nor isoflavone-rich soy affect total or free testosterone (T) levels."

Lawrence Ross, MD, is a professor at the University of Illinois Medical School whose specialty is male infertility and microsurgery. Dr. Ross does not believe that there is any actual data to support the claim that soy has a feminizing effect on men. According to Dr. Ross-

> There are, to my knowledge until the present time, no significant trials of soy saying what level might cause that problem or a level of soy that actually does cause the problem. So it is a theoretical concern, but to date there is no good literature, no good studies to prove that it's detrimental.
>
> There are populations that consume large amounts of soy-based products, for instance in China, and there is no evidence at least to date that has

> been presented that say that Chinese men have lower sperm counts related to soy-based diet than men in the US or Europe or anywhere else.

Regarding the anti-soy rhetoric published on the Internet and elsewhere referencing in vivo (animal) studies, Dr. Ross had this to say:

> The Internet is not a source of peer-reviewed science. Anybody who wants to say anything, interpret anything, can put it on the Internet and just as we say about computer function you know, "garbage in, garbage out."
>
> The problem with trying to quote animal studies, I can take you back thirty years when artificial sweeteners hit the market. There were several studies in mice that showed that when you gave mice very large quantities of artificial sweeteners, they developed bladder cancer.
>
> That was a valid study; however, the amount of artificial sweetener that was fed to mice, to induce bladder cancer in mice, was beyond a quantity that any reasonable human could ever ingest or would ever ingest.
>
> I think for those of us who see fertility patients and are asked these questions, who ask if soy is safe to eat, I say that as far as I can tell, from scientific studies that are valid studies out there, that there is no evidence that the moderate ingestion of soy, and then you can argue about what's moderate and what isn't, has no deleterious effect on fertility—at least we have no evidence of that effect.

Nutrition researcher and Physicians Committee for Responsible Medicine president, Dr. Neal Barnard comments:

> Soy products have no effect on testosterone levels. It doesn't have any effect on infertility at all, one way or another. And if it did we wouldn't be seeing the explosion of populations throughout Asia, where soy is very widely consumed. It obviously does not cause any problems whatsoever with regard to malc sexual functioning
>
> Let me be clear, researchers have looked very carefully at diet and testosterone levels in men and the evidence is clear that soy products don't cause any problems whatsoever with regard to testosterone levels or male sexual functioning.

Regarding soy consumption and infertility, Dr. Ross states,

> Soy is a sort of a phytoestrogenic compound. So presumably, if it were ingested in high enough quantities to have an estrogenic effect, that would affect pituitary function and testosterone output that would be a theoretical concern. But again, I am not aware of anyone showing that moderate consumption of soy can actually cause those changes. Right now, there is nothing in the literature to suggest that.

When asked about the feminizing effects that phytoestrogens in soy purportedly cause, Dr. Messina replies, "The isoflavones in soyfoods, which are classified as phytoestrogens, have raised concerns about feminization. However, isoflavones are different from the hormone estrogen. Dr. Messina continues:

> Studies show quite clearly that the two molecules, isoflavones and estrogen, exert different physiological effects. For example, estrogen raises levels of tri-

glycerides, which increases risk of heart disease, and raises levels of HDL-cholesterol, which lowers risk. In contrast, isoflavones have no effect on triglycerides or HDL.

Nor do isoflavones have any detrimental effects on male hormones. A meta-analysis examined the relationship between soy/isoflavone intake and reproductive hormone levels in men. The study included thirty-six treatment groups and found no effects on total and free testosterone levels. By the way, three clinical studies also show no effects on sperm or semen.

Consuming soy may actually help prevent sex organ cancers. A recent report in the *American Journal of Clinical Nutrition* analyzed fourteen separate studies and showed a 26 percent reduction in the risk for prostate cancer among those who consumed soy.[7]

Soy has been considered a beneficial dietary component among Asian populations for centuries; however, negative suppositions in the press and elsewhere prompted an international contingent of scientists to conduct a review. The researchers examined more than two hundred studies and wrote in their conclusion, "The available scientific evidence supports the safety of isoflavones as typically consumed in diets based on soy, or containing soy products."[8]

CHAPTER 9

Rumor: Eating Tofu Causes Alzheimer's Disease

Anti-soy claim:
Consuming tofu shrinks the brain and causes Alzheimer's and dementia.

Alzheimer's disease is the most common form of dementia and is most often diagnosed in people over the age of sixty-five; however, it can develop earlier. Alzheimer's is a progressive form of senile dementia. It damages areas of the brain involved in memory, intelligence, language, judgment, and behavior.

In November of 1999, a headline in the *Honolulu Star-Bulletin* Hawaii News section screamed, "Too much tofu induces 'brain aging,' study shows," with the sub-heading, "A Hawaii research team says high consumption of the soy product by a group of men lowered mental abilities." The story was picked up and ignited a firestorm of controversy.

The Honolulu-Asia Aging Study (HAAS),[1] published in 2000, reported a shocking and unprecedented finding that was very unsettling to legions of health conscious consumers. Headlines such as, "Tofu Shrinks the Brain," spread like wildfire and associated midlife tofu consumption with poor cognitive test performance, enlargement of ventricles, and low brain weight. The research cited was from the Honolulu Heart Program, an ongoing study of the health of Japanese-American men living in Hawaii.

The study, whose lead investigator was Dr. Lon White, was part of the Honolulu Heart Study, which was investigating the diet and risk of dementia of Japanese men residing in Hawaii.

Dr. White and his associates' surprising findings reported that those men who consumed the most tofu during their mid-forties to mid-sixties showed the most signs of mental deterioration and were more likely to have developed dementia and Alzheimer's as they grew older, with onset in their seventies to early nineties. High tofu consumption was defined in this study as being consumed two or more times a week.

The study drew a strong correlation between tofu and cognitive decline independent of other factors like age, education, or obesity.

The Honolulu-Asia Aging Study raised many questions. The Japanese lifestyle has long been associated with better cognition in old age and the rate of dementia in Asia is considered to be much lower than in western countries. If tofu consumption increased the incidence of Alzheimer's disease, then there would be more Alzheimer's in Japan than in Hawaii, because more tofu is consumed in Japan. But in fact the reverse is true.

Dr. White's findings were especially surprising in light of the fact that soy protein has been shown to reduce blood cholesterol dramatically and high levels of cholesterol have been associated with an increased risk for Alzheimer's disease. The FDA approved the health claim recommending twenty-five grams of soy protein per day to reduce the risk of coronary heart disease (CHD) in 1999.[2]

According to Dr. Mark Messina, there is a possible biological explanation for the findings.[3]

> Soybeans contain isoflavones, which are weak estrogens. They fall into the category of estrogen-like compounds known as SERMS—selective estrogen receptor modulators.[4]
>
> This means that they have estrogenic effects in some tissues and anti-estrogenic effects in others. Estrogen may have a positive effect on brain tissue but the researchers of the Hawaii study suggested that isoflavones may have anti-estrogenic effects on the brain.
>
> Of course, we can't know this from the Hawaii study. This was an epidemiological study, so it doesn't show cause and effect. It merely shows that two things occur together. Since the researchers measured intake of only twenty-seven foods and were not able to control for every single lifestyle factor, it is possible that tofu consumption is a marker for some other factor

> that affects cognitive function. This would make tofu an innocent bystander.

It is important to note that the Honolulu Heart Study is the only study that has suggested that there is a link between tofu consumption and dementia. Dr. Messina states:

> One study has suggested a link between tofu consumption and poorer cognitive function in old age, but this is an epidemiological study. Therefore it doesn't show cause and effect.
>
> It did not look at diet extensively enough to draw firm conclusions. There are no other studies to support it and clinical studies suggest soy and isoflavones have beneficial effects on cognition.

Dr. Messina concluded, "At this point, there is no reason to believe that eating soyfoods is harmful to brain aging."

Epidemiology is the study of patterns of health and illness and the associated factors in populations and is a method used most often in public health research to identify risk. The most popular among epidemiological methods is the observational study. An observational study is one in which a group of individuals are observed or certain outcomes are measured. In this kind of study, independent data is statistically analyzed and evaluated.

The findings with regard to this paper concerned a particular group of Asian men who were living in Hawaii and who consumed tofu. The researchers focused on the consumption of tofu exclusively, so that they did not include other forms of soyfoods—such as tempeh, soymilk, or anything like that—it was specifically tofu.

The HAAS study was met with a fair amount of skepticism. Dr. Carey Gleason, a Senior Scientist in the field of Geriatrics and Gerontology at the University of Wisconsin-Madison, points out several flaws in the HAAS study.

> First of all, it was an observational study and it was criticized for a couple of different reasons, and I think the authors of the study would actually recognize that their paper might have been misinterpreted by the lay public.
>
> What they did is they controlled for certain variables—what we call "confounding variables" or confounders.[94] In this type of research . . . it's just the nature of the research that it's an imperfect way of trying to account for the variables by actually correcting for them afterwards, not controlling for them.

Randomized clinical trials (RCT) are a common type of scientific experiment whereby individuals with similar traits are assigned to different groups and the randomization controls are set in place at the outset.

In the RCT model, randomization is in place to try to control for variables by randomly assigning people to different groups. However, in the observational model, you cannot randomly assign people as to whether or not they've eaten tofu and the scientists are basically trying to retrospectively correct that imbalance, which is a much weaker statistical model for which no firm conclusions can be drawn.

What the researchers in the HAAS study know is that the people who ate more tofu happened to be older men who were less acculturated.

So, these men may have been more likely to be born in Japan, as opposed to younger men who were born in Hawaii. It is then more likely that there may be other things that could have accounted for the smaller brain size.

It could be that these men are less likely to use Western medicine and there may have been educational differences or perhaps their diet as a child was poor. Any or all of these factors could be in play.

Like any analytic methodology, epidemiology has its hazards. Dr. Gleason explains:

> In these observational studies, you are trying to correct for cohort differences by adjusting in the statistical models, which is never going to be as good as controlling for them at the outset by randomization.
>
> That study did have that fatal flaw that I think the authors themselves would recognize and not want their results to be misstated. They simply wanted to describe the fact that in people who consume more soy, there seemed to have smaller brain size.

As is often the case with the anti-soy claims, ongoing studies are hijacked midstream and broadcast on the Internet as a new scientific discovery. According to Dr. Gleason:

> The flaw with epidemiological studies such as this one is they are trying to pick out one of a thousand different elements in a person's diet and associate that with a kind of global outcome, cognition or brain size. In order to do that you need large samples and you are trying to adjust models for what you think are other factors influencing the outcome.

There is much research with results that suggest the opposite of those in the Honolulu Heart Study.

Seventh Day Adventists are vegetarian, many of whom consume soyfoods their entire lives. They avoid meat and consume a diet of legumes, whole grains, nuts, fruits, and vegetables and have a long history and involvement with soyfoods.

According to the SoyInfo Center SoyaScan database, no other organization or group of people has played a more significant role than Seventh-day Adventists in introducing soyfoods, vegetarianism, and other meat alternatives to the Western world.

In a prospective study of Seventh-day Adventists, participants were matched for age, sex, and zip code. Those who ate meat (including poultry and fish) were more than twice as likely to become demented as their vegetarian counterparts.[6]

Clinical studies support the evidence that cognitive function actually benefits from a diet rich in soy. Researchers at Loma Linda University in California conducted a study with Seventh Day Adventists over the age of seventy-five.[7] Researchers found that these elderly subjects had a lower incidence of dementia in old age than the general population.

One such study evaluated young adult men and women who consumed a diet high in soy and experienced substantial improvements in short-term and long-term memory as well as mental agility.[8] Other studies have found that isoflavone supplements from soy improve cognitive function in postmenopausal women.[9] [10]

The studies most often cited with regard to cognition and diet are those conducted among Seventh-day Adventists, who are vegetarian and as a rule consume soy alternative products regularly.

Seventh-day Adventists (SDA), from the age of sixty-five and over who had not eaten meat in the previous thirty years were found to be about one-third as likely to develop dementia as their regular meat-eating counterparts.[11] Another study of California Seventh-day Adventists, where the actual soy intake compared to dementia rates was not the primary focus, showed the SDA vegetarians to eat an average of 3.5 servings of meat substitutes (which usually contain soy) per week compared to only 1.4 servings by their meat-eating counterparts.[12]

The Honolulu Aging Study is the only study that has suggested a link between tofu consumption and dementia in old age. Soy researchers Mark and Virginia Messina have concluded, "There is no reason to believe that eating soyfoods is harmful to brain aging."

CHAPTER 10

Rumor: Eating Soy Causes Cancer

Anti-soy claim:
The phytoestrogens and other substances in soy cause cancer of the breast, pancreas, and prostate.

Soy has long been recognized as a nutrient-dense food with many anti-cancer properties. The cancer protective effect of a soy-based diet has been the subject of numerous studies and researchers have been investigating which constituents in the soybean are responsible.

One such study recently found strong anti-cancer properties using soybean meal from a variety of soybean lines high in oleic acid and protein. The researchers observed bioactivity between the peptides derived from the soybeans and various types of human cancer cells.[1]

The study showed that peptides derived from soybean meal significantly inhibited cell growth by 73 percent for colon cancer, 70 percent for liver cancer, and 68 percent for lung cancer cells using human cell lines. This demonstrated that the selected high oleic acid soybean lines could have a potential nutraceutical affect in helping to reduce the growth of several types of cancer cells.

The interest in soy and cancer prevention stems from its ability to impede tumor growth. Researchers specializing in investigating the relationship between soyfood consumption and cancer prevention at the National Cancer Institute found soyfoods to be rich in anti-carcinogens. These compounds in soybeans may help prevent various forms of cancer in a number of ways.

Five different chemical classes of anti-carcinogens are present in soy: phytoestroles, phytates, saponins, protease inhibitors, and isoflavones. Soybeans are also rich in phenolic acids, many of which contain anti-carcinogens.

Soy isoflavones have been shown to induce cellular differentiation. Differentiation is the process that defines whether a cell will develop into a nerve cell, heart cell, or a kidney cell. Cancer tumors contain undifferentiated cells and soy isoflavones have been shown to encourage cancer cells to differentiate into normal cells.

Research suggests that the isoflavones in soy may inhibit cellular proliferation, arresting the spread of cancer by blocking an enzyme called tyrosine protein kinase, which cancer cells employ during their accelerated growth.

Additionally, studies have also shown that the isoflavones in soy can actually reverse angiogenesis, which is the growth of blood vessels that feed tumors and allow them to grow.[2]

The two most studied isoflavones, genistein and daidzein, are the phytoestrogens that have garnered the most press on both sides of the issue.

Isoflavones are weak plant estrogens that are thought to help protect against breast, ovarian, and prostate cancer by inhibiting the growth of hormone-dependent tumors.

High blood levels of estrogen are a well-established risk factor for breast cancer and it has been hypothesized that these weak plant estrogens compete for the cell receptors and thus are protective against this type of cancer.

Observational studies of populations where the consumption of soy products is high support this hypothesis. For example, the Japanese breast cancer mortality rate is only one quarter of that in the United States.

In 2008, Chinese researchers showed that genistein inhibits the proliferation and activation of the enzyme tyrosine protein kinase (TPK), which is the enzyme involved in controlling cell growth and regulation.[3]

Numerous studies and publications suggest that soy isoflavones have a protective effect against menopausal symptoms, osteoporosis, heart problems, cancer and more. However, saponins found in soybeans are now thought by many researchers to have important synergistic functions with isoflavones.

Saponins, classified as polyphenols, are one of the main categories of plant compounds and are more abundant in soybeans than in any other legume. Soy is well known to reduce cholesterol and saponins have been shown to reduce total cholesterol in animal models. One of the most exciting prospects for saponins is how they appear to inhibit or kill cancer cells without killing normal cells, unlike many pharmaceutical drugs, which have serious side effects.

Diet is a major factor in the development of colon cancer. It has been reported that the colon cancer mortalities are higher in countries where people consume more animal products.[4]

Soy saponins have been found to have inhibitory effects in vitro on cancer cells by changing the cell structure, cell proliferation enzymes, and cell growth.[5]

Prostate Cancer

Prostate cancer is the second leading cause of cancer-related death in American men. The prostate is a gland in the male reproductive system and cancer cells can metastasize (spread) from the prostate to other parts of the body, particularly the bones and lymph nodes.

Research has shown that soy isoflavones have anti-carcinogenic effects against prostate cancer. Epidemiological and experimental evidence suggests that an increased consumption of soy may reduce the risk for developing prostate cancer.

One such study found that food products, which contain a combination of active compounds, may be safer and more effective in fighting cancer than individual compounds.[6]

Researchers believe that a whole-food based approach is significant for the development of public health recommendations for prostate cancer prevention.

According to a study published in the *Journal of the National Cancer Institute*, genestein, a compound found in soybeans, almost completely prevented the spread of human prostate cancer in mice.[7]

In this study conducted at Northwestern University Department of Medicine in Chicago, research scientists investigated the target for genistein in prostate cancer cells. They assessed cell invasion and the gene and protein expression of mitogen-activated protein kinase 4 (MEK4) and matrix metalloproteinase-2 (MMP-2), which is associated with cell invasion.

The authors write:

> We have shown that it is possible to target motility-associated processes with genistein in patients with prostate cancer, have identified MEK4 as the therapeutic target for genistein in all six prostate cell lines examined and have provided a possible mechanism to link high dietary consumption of genistein-containing foods with lower rates of prostate cancer metastasis and mortality.

A recent Chinese study analyzed isoflavone levels in the blood and discovered that those who had consumed soy had the lowest likelihood of developing prostate cancer, and further, that soy greatly reduced the risk of the cancer metastasizing, or spreading throughout the body.[8]

In another recent finding, researchers at Ohio State University found that soy decreases multiple inflammatory markers that can impact prostate cancer and progression, including reducing infections. Early stage prostate cancer patients had 68 milligrams soy isoflavones daily for eight weeks via soy flour bread. Soy intake led to an improved immune system response, reduced inflammation related to early development of cancer cells, and encouraged an active immune system throughout cancer progression.[9]

Breast Cancer

Estrogen is a known risk factor in the development of breast cancer. Two studies suggested that the isoflavones in soy may actually stimulate the growth of estrogen-dependent breast cancer tumors. This was a conclusion that had not been reached by researchers previously and whether cancer patients should consume soy became the subject of some controversy.

Concern over the possibility that consuming soy could actually stimulate the growth of tumors has caused a good deal of confusion among oncologists and other health care professionals.

This issue has been frustrating for breast cancer patients and has caused many women to view soy with fear and anxiety. This turn of events has been particularly ironic, because a great deal of research showing the anti-carcinogenic effects of soy has been the source of excitement for decades.

I spoke with Stacey Krawczyk, MS RD, of the National Soybean Research Laboratory in Urbana, Illinois about whether there is anything to the rumor that consuming soyfoods could be harmful and even cause cancer.

> What I struggle with the most is anytime soy is reported, especially in the media, and often the researchers talk about soy; however, soy to them can mean whatever they're looking at. People who are looking at the research are looking at isolated isoflavones.
>
> Results of research done with isolated isoflavones, often in doses administered in huge quantities, are regularly reported in the media as "soy," which the general public thinks of as soyfoods.
>
> Unfortunately, there seem to be some very strident anti-soy voices out there. They are very media savvy and they have become very good at sound bites and at repeating those sound bites. When you repeat something often enough people believe it, whether or not it is true.

Speaking to the extracted components of soy that are commonly used in very high concentrations in so much of the research, Stacey had this to say, "I think that valid research illustrates that isolated components are generally harmful to the body and in this case, isoflavones, especially related to breast cancer."

"However, what is reported is about soy. . . and it is not soy. When I think of soy, I think of soyfoods. Once you start pulling out and teasing out things individualized, then you have more risk in the body."

When you isolate a component and consume it in large amounts, what are the potential ramifications for your body? Whether that be taking an isoflavone supplement or a bushel of carrots? If you are artificially consuming something at levels beyond what exists in nature, what impact does that have on the body?

According to dietician Stacey Krawczyk, "With regard to actual foods, that is where the Shanghai study showed that consuming soyfoods is actually prevention. That for women, even women on Tamoxifen and breast cancer survivors, eating soyfoods helps to reduce their risk of recurrence."

The New Puberty, a recent book by Louise Greenspan, MD, and Julianna Deardorff, PhD, looks at the percentage of girls who are now going through early puberty and the environmental, biological, and socioeconomic factors that could be the cause of this phenomenon.[10]

One of the authors, Dr. Louise Greenspan, a clinical pediatric endocrinologist, addressed the impact of soy on breast cancer in a recent interview, noting that "soy is actually protective and that higher soy intake may lead to later puberty," because the estrogen-mimicking effects of soy may "down-regulate the estrogen receptor, so that later in life, your body doesn't perceive or see estrogen in quite the same way."

Dr. Leena A. Hilakivi-Clarke is a professor of Oncology at Georgetown University School of Medicine and an American Institute for Cancer Research-funded scientist. Dr. Hilakivi-Clarke's work includes the investigation of how dietary exposures to estrogens in utero or during childhood can have an impact on breast cancer tumors later in life.

Dr. Hilakivi-Clarke's work suggests that consuming soy earlier in life and even exposure to soy in the womb may exert an extra measure of cancer protection.

I asked Dr. Hilakivi-Clarke about women consuming isoflavone supplements as opposed to eating actual soyfoods.

"No, I don't think women should be taking isoflavone supplements. The whole food is fine, preferably the typical Asian soyfoods. If you're taking the isoflavone supplements, you are exposing yourself to the isoflavones at a very high level."

Dr. Hilakivi-Clarke referenced the work of William Helferich at the University of Illinois. "He has been doing lots of studies in models that mimic postmenopausal women. He found that if you give just purified genestein,[11] you get an increase of human breast cancer cells growing in nude mice. But, if you give it in the context of soya flour, then it did not."

According to the *Journal of Family Practice*, high dietary soy intake does not affect a woman's risk of primary or recurrent breast cancer. However, it does (favorably) affect the risk of cancer recurrence. "Compared with diets low in soy, high dietary intake of soy protein or soy isoflavones isn't associated with any alteration in the risk of developing primary breast cancer."[12]

The American Institute of Cancer Research has this to say about the issue:

> Soyfoods are good sources of protein, and many are also good sources of fiber, potassium, magnesium, copper, and manganese. Soy contains a variety of phytochemicals and active compounds, and because soy contains estrogen-like compounds, there was fear that soy may raise risk of hormone-related cancers. Evidence shows this is not true.

Eating Soy Boosts Breast Cancer Survival

"Soyfood intake and breast cancer survival," published in the *Journal of The American Medical Association* (JAMA), analyzed data from the Shanghai Breast Cancer Survival Study, the largest population-based study of breast cancer survival to date. The Shanghai study included 5,042 women from twenty to seventy-five years of age and followed them for a period of four years.

Soy consumption for cancer patients had been the subject of some controversy, and this 2009 study showed that breast cancer survivors who consume soyfoods reap important health benefits and found that eating soyfoods can increase the rate of survival for breast cancer patients.

According to the report published in JAMA, women diagnosed with breast cancer and who consume soyfoods such as tofu, soymilk, or edamame reduce their risk of recurrence by 32 percent.[13]

Previous Research Refuted

Previous research had produced contradictory results with some studies suggesting that soyfoods reduce the risk of breast cancer, while two others suggesting that compounds unique to soy may help breast cancer cells to grow.

Studies have indicated that the relative levels of genistein and estrogen at the target site are important to determine the genistein effect on the ER-positive tumor growth. However, studies using ovariectomized mice and subcutaneous xenograft models might not truly reflect estrogen concentrations in human breast tumors.

Researchers recommend that in future preclinical studies, the relative levels of isoflavones and estrogen at the target site should be considered more carefully in the study design, as well as different ER-positive breast cancer cells.

Findings obtained in two recent human studies show that a moderate consumption of isoflavone does not increase the risk of breast cancer recurrence in Western women, and Asian breast cancer survivors exhibit better prognosis if they continue consuming a soy diet.[14] [15]

Now, previous theories have been refuted by studies demonstrating that soy does not increase the growth of breast cancer cells and has been shown to increase survival rates.

Higher Soy Intake, Lower Mortality

Researchers also found that breast cancer patients who consume soy had a 29 percent decreased risk of death, as compared to women who consume little or no soy.[16]

Xiao Ou Shu, MD, PhD, lead researcher and professor of medicine at Vanderbilt University Medical Center in Nashville said, "Women who had a higher soy intake had a lower mortality and lower risk of relapse."

Soy Compounds May Reduce Estrogen in the Body

Soybeans are rich in phytoestrogens, which are also known as isoflavones. Although these substances are one thousand times less potent than human estrogen, there had been some concerns that isoflavones may have an estrogen-like effect and may increase cancer risk.

There are many ways that phytoestrogens may work in the body. While phytoestrogens have effects that are different from those of estrogen, the chemical structure of phytoestrogens is similar to estrogen.[17]

Estrogens hook onto tiny receptor proteins in cells that allow them to change the cell's chemistry. Phytoestrogens may act as mimics of estrogen. When phytoestrogens occupy the cell, normal estrogens cannot. Plant estrogens do not eliminate all of estrogen's effects, but they do minimize them, apparently reducing breast cancer risk and menstrual symptoms.[18]

Soy is Safe and Potentially Beneficial

An editorial accompanying the "Soyfood intake and breast cancer survival study"[19] suggests some probable inconsistencies in prior research. The editorial attributed those inconsistencies to the fact that soy consumption in the US is a good deal lower, which made the beneficial effects of consuming soyfoods difficult to identify.

In China, soy intake is higher and diets tend to include the intake of more traditional soy from food sources, rather than from soy supplements.

The researchers report:

> The inverse association was evident among women with either estrogen receptor-positive or receptor-neg-

> ative breast cancer and was present in both users and nonusers of tamoxifen.
>
> In summary, in this population-based prospective study, we found that soyfood intake is safe and was associated with lower mortality and recurrence among breast cancer patients.

These scientists concluded that, among women with breast cancer, soyfood consumption was significantly associated with decreased risk of death and recurrence. Dr. Shu and her colleagues stated, "This study suggests that moderate soyfood intake is safe."

I asked Stacey if, in her role as nutrition professional, she would recommend consuming soy to cancer patients and survivors. She responded:

> My mom is a breast cancer survivor. She's on Tamoxifen right now and she has had a hysterectomy. She was on estrogen replacement therapy for many years prior to her diagnosis for breast cancer.
>
> I tell her that including soyfoods in her diet is a great thing for her to do and that it is an excellent source of protein and is low fat, low saturated fat, and contains no cholesterol.
>
> My mom has high cholesterol and she's overweight, so soyfoods present a great opportunity for her. However, I would not tell her to consume an isoflavone supplement.
>
> This is a topic that is widely misreported in the media where isolated isoflavone vs. soyfoods is watered down to say that "soy is bad." I think that when you look at a much higher consumption than normal we have to be cautious.
>
> It would be difficult to over consume soyfoods as opposed to isoflavone supplements, because food takes

up volume and there is going to be a natural barrier to over consume and the stomach can only hold so much.

The incidence and mortality of breast cancer varies widely around the world. A number of studies tracing the migration of people from one culture to another suggest that lifestyle factors are the principal influence. The high soy intake among Asians has been associated with an almost one-third reduction in the risk of breast cancer.

In a recent study published in the *Journal of Clinical Nutrition*, Dr. Mark Messina and Dr. Anna Wu investigated the relationship between breast cancer and soy.[20] The authors write:

> The ability of the isoflavone, genistein, to stimulate the growth of mammary tumors in ovariectomized (ovaries were surgically removed) athymic nude mice (without an immune system) implanted with estrogen-sensitive breast cancer cells has raised concern that soyfoods and especially isoflavone supplements are contraindicated for patients with breast cancer and women at high risk of breast cancer.
>
> The concern that products containing isoflavone might be contraindicated for patients with breast cancer and women at increased risk of breast cancer is based almost exclusively on results from rodent studies.
>
> In contrast, however, the clinical and epidemiologic data suggest that isoflavones pose no risk to such women. This suggestion is consistent with the relatively unimpressive data showing that postmenopausal therapy with oral estrogen increases breast cancer risk.

These researchers noted that the relation between soy and breast cancer has been studied for more than twenty years and found the epidemiologic data indicates an association between the high intake of soy among Asians and a decreased incidence of breast cancer.

A study published in the Journal of Biochemistry and Molecular Biology suggests that genestein could be of therapeutic value in preventing human breast cancer.[21]

The isoflavone, genistein, has been shown to inhibit the growth of both estrogen-dependent and estrogen-independent breast cancer cells.[22] The growth of estrogen-dependent or estrogen receptor–positive breast cancer cells is fueled in a high estrogen environment.

In this study, Korean scientists investigated the chemopreventive (anti-carcinogenic) and cytotoxic (cell toxicity) effect of genistein against human breast cancer cell lines. Genistein-inhibited cell proliferation in both estrogen receptor–positive and estrogen receptor–negative human breast carcinoma cell lines.

The JAMA report on the Shanghai Breast Cancer Survival Study clearly refuted any questions raised in animal models.

The Shanghai study analyzed actual women diagnosed with breast cancer who consume soyfoods such as tofu, soymilk, or edamame who actually reduced their risk of recurrence by 32 percent.[23]

This study with human patients consuming soyfoods, not supplements, has disproved previous theories based on rodent models by demonstrating that soy did not increase the growth of breast cancer cells and crediting soy with an increase in survival rates.

In China, soy intake is higher and diets tend to include the consumption of more traditional soy from food sources, rather than from soy supplements, which are more popular in the US.

The editorial goes on to say:

> Even though the findings by Shu et al. suggest that consumption of soyfoods among breast cancer patients is probably safe, studies in larger cohorts are required to understand the effects of these foods among diverse clinical subgroups of breast cancer patients and survivors.
>
> In the meantime, clinicians can advise their patients with breast cancer that soyfoods are safe to

> eat and that these foods may offer some protective benefit for long-term health.[24]

The researchers report, "The inverse association was evident among women with either estrogen receptor-positive or receptor-negative breast cancer and was present in both users and nonusers of tamoxifen."

Tamoxifen is an oral drug used in the treatment of breast cancer and other types of cancer that block or interrupt the hormone estrogen. Tamoxifen has been shown to slow or stop the growth of cancer cells presenting the body.

Like all pharmaceutical drugs, Tamoxifen has a list of side effects that include blood clots, especially in the lungs and the legs,[25] strokes,[26] endometrial and uterine cancer[27], and cataracts,[28] as well as hot flashes and vaginal discharge. Some women experience irregular menstrual periods, headaches, fatigue, nausea, vomiting, weight gain, vaginal dryness or itching, irritation of the skin around the vagina, and skin rash.[29]

"In summary, in this population-based prospective study, we found that soyfood intake is safe and was associated with lower mortality and recurrence among breast cancer patients."

These scientists concluded that, among women with breast cancer, soyfood consumption was significantly associated with decreased risk of death and recurrence.

Dr. Dixie Mills is a Harvard-trained surgeon who has specialized in breast care since 1989 and believes that soyfoods can be a healthful addition to a balanced diet. "I usually talk about nutrition with my patients and I certainly support women wanting to make changes in their diet to include soy." According to Dr. Mills:

> Phytoestrogens don't disrupt endocrine function; they collaborate with it. I think soyfoods could be cancer-protective. It's looking like it from the soy studies I've seen that it is protective.

> The anti-soy people make soy this awful poison. They say, "Oh, soy interferes with Tamoxifen" and make it powerful in that way, but they don't make it so powerful that it could protect.

These well-documented, peer-reviewed studies have not deterred the soy antagonists from ringing the bell with alarmist tactics, proclaiming that soy causes pancreatic cancer and other cancers, including infantile leukemia.[30]

The rhetoric always circles back to the same few sources, Sally Fallon and Mary Enig, and others of the WAPF, Dr. Joseph Mercola, and Richard and Valerie James of New Zealand. These self-proclaimed soy antagonists pepper their articles liberally with serious implications accusing highly respected researchers and government officials of sinister behavior and outright bias.

Fallon and Enig claim that the cancer-causing effects of soy have been ignored, charging that a highly respected soy research scientist, "conveniently neglected to include" a particular study in his meta-analysis.

Fallon and Enig draw their conclusion regarding the relationship between soy and cancer by stating that feeding with soy caused pancreatic cancer in one animal study. However, the study referenced would appear to be about isolated trypsin inhibitor and not at all about feeding with soy.[31]

Whole Food vs. Piecemeal Approach to Evaluating Soy

As we have seen, the research cited by the soy antagonists is based on animal studies, where massive quantities of isolated compounds were introduced into the body.

Despite the fact that the health benefits enjoyed by regularly consuming soyfoods among various populations is quite clear, soy researchers continue to isolate various elements of the in search of identifying its healthy components. Sadly, investigating the benefits of whole soy remains underrepresented in the scientific community.

In a paper published in late 2010 by the Department of Food and Nutrition at Perdue University, "whole versus the piecemeal approach to evaluating soy,"[32] investigators suggest that the well-known benefits attributable to soy have not been consistently proven in Western populations who oftentimes opt to consume more processed soyfoods, protein powders, or supplements.

"Soy has been singled out for attention among other legumes as a valuable source of nutrients, phytochemicals, and bioactive compounds. Early epidemiological studies established that whole soy and traditional soyfoods were implicated in health-protective effects in Asian populations."

Principal investigator Dr. Reinwald and colleagues wrote:

> Various dietary guidelines advocate the regular consumption of legumes that tend not to be included in our diets in sufficient quantities. This paper highlights the possibility that whole soy may have a more unique effect on health than a select soy component(s).

The researchers explored the rationale for focusing research on whole soy in an attempt to understand it better rather than trying to replicate the health benefits by targeting various soy extracts.

The authors conclude, "Soybeans are a good source of bone-healthy nutrients. Epidemiological studies in Asia evaluating diets containing traditional whole soyfoods show a positive association with bone mineral density and fracture protection."

The researchers note that smaller scale intervention studies in Western countries feature isolated soy protein and purified or concentrated soy isoflavones rather than whole soyfoods and that they have produced inconsistent results.

CHAPTER 11

Rumor: Soy Causes Bone Deficiencies

"Soyfoods can cause deficiencies in calcium and vitamin D, both needed for healthy bones. Calcium from bone broths and vitamin D from seafood, lard and organ meats prevent osteoporosis in Asian countries-not soyfoods."[1]

—Weston A. Price Foundation brochure

Osteoporosis is a major concern, particularly for postmenopausal women. Osteoporosis—literally "porous bone"— is the progressive degeneration of bone mineral density and bone strength, which results in thinner, more porous bones and increases bone fragility making them susceptible to malformation and fractures. While osteoporosis can develop most anywhere in the bone structure, most often bone loss occurs in the hips, ribs, and spine.

It is the drop in levels of estrogen associated with menopause that puts women at an increased risk for decreasing bone density and fractures. Bone loss occurs most rapidly after menopause when women may lose up to 20 percent of bone mass in the five years following menopause. Older men are not immune to the condition and can also develop osteoporosis as their hormone levels decrease.

While estrogen is known to increase bone mass, estrogen replacement therapy is associated with adverse effects; such as increased risk of endometrial cancer and breast cancer and many women would prefer alternatives to estrogen.[2]

A number of studies indicate that there are compounds in soy that benefit bone health. One such study, published in the Archives of Internal Medicine, showed that soyfood intake was associated with a significantly lower risk of bone fracture, particularly among early postmenopausal women. [3]

A recent study by Purdue University found that daily soyfood consumption was very beneficial to bone retention for postmenopausal women. Researchers report that 105 milligrams of isoflavones a day (about four servings of traditional soyfoods over a fifty-day period) increased bone calcium retention by 7.6 percent among the women. Soy consumption was shown to be an effective bone-preserving agent not only at that level, but was also beneficial even when lower amounts of soyfoods were consumed.[4]

Unlike the preponderance of research, which works with extracted compounds, these scientists evaluated actual soyfood consumption. The research team examined the relationship between normal soyfood intake and fracture incidence in 24,403 postmenopausal women who had no history of fracture or cancer over a period of four and a half years.

Dr. Marta D. Van Loan is with the USDA Agricultural Research Center Western Human Nutrition Research Center at the University of California, Davis. The Western Human Nutrition Research Center is involved with creating and testing nutrition interventions to improve health and assessing how an individual's environment and genetics affect nutrient metabolism. They are also charged with providing reliable and reproducible research results for the development of national nutrition policies.

I asked Dr. Van Loan about the claim that soy could cause bone deficiencies and even precipitate osteoporosis, as claimed by the WAPF. Dr. Van Loan explained:

> It may be that people are deficient because they don't consume enough calcium and vitamin D, but that's not to say that soy caused it. Those are two entirely different issues, so that claim is a bit out there.
>
> Ever since conventional steroid hormone replacement therapy was shown to cause certain kinds of cancer and other side effects, researchers have been looking for a safe and effective alternative for postmenopausal women.
>
> Soy is one potential candidate and has been the subject of more than two dozen studies conducted in the US and abroad during the past decade. Some of those investigations have suggested that soy enhances bone health.

In a meta-analysis conducted by a team from Peking University in Beijing, China and the University of Yamanashi in Japan, soy isoflavones were found to prevent bone resorption (the break down of bone resulting in a transfer of calcium to the blood), increase bone mineral density, and stimulate bone formation in menopausal women.[5]

Results indicate that, in addition to inhibiting bone resorption and increasing bone formation, soy was also found to improve spinal bone mineral density.

Studies show that, in spite of their significantly weaker estrogen action, isoflavones have bone-building effects. It is widely accepted that this is the reason why osteoporosis is very rare after menopause in Asian countries, despite their low consumption of dairy products.

In 2004, it was demonstrated that soy isoflavones have a beneficial effect on bone mineral content, especially in women who were over four years into menopause and had either a lower body weight or a low calcium intake.[6]

In the human body there is a constant process of breaking down and remaking of bones. A 2011 meta-analysis reviewed five studies regarding the effects of isoflavones on bone mineral density and bone turnover. They concluded that soy isoflavones significantly increased the bone mineral density of the lumbar spine and moderately decreased certain bone resorption markers. They concluded that soy isoflavones may not only prevent osteoporosis, but also improve bone strength in postmenopausal women.

Calcium is the most abundant mineral in the body and is found in some foods, added to others and available as a dietary supplement. We need calcium in small amounts for the growth and formation of the bones, blood clotting, and nerve and muscle function.

The role of calcium intake and the onset of osteoporosis has been the theme of numerous advertising campaigns in recent years, supported by mainstream medical advice. This has caused a general fear that our diet is somehow lacking in adequate calcium.

If you believe the dairy industry, cow milk, and the myriad of dairy products made from it are an essential part of a healthy diet. Many medical and nutrition professionals disagree.

Some experts believe that the idea that osteoporosis is caused by calcium deficiency was created to sell dairy products. The relentless propaganda of the dairy industry, which has been manipulating public opinion since the early 1950s, is possibly one of the best advertising schemes in modern food industry history.

The USDA recommends three glasses of milk a day, even though significant studies show that there is no positive association between milk and bone strength. The government has been complicit in the promotion of dairy products to the degree where "Got Milk" has become a national catch phrase. One of the most insidious ways the agency has endorsed dairy products is by placing them prominently in the USDA Food Guide Pyramid.

Created in 1862, the mission statement of the USDA is to enhance the quality of life for Americans by supporting the production of agriculture. In modern-day America, the USDA is charged with assisting dairy farmers while promoting healthy dietary choices for consumers. This creates a conflict of interest that puts the objectivity of government farm policy and the health of the dairy-consuming public at risk.

In December 1999, the Physicians Committee for Responsible Medicine (PCRM) filed suit against the USDA, claiming that the department unfairly promotes the special interests of the meat and dairy industries through its official dietary guidelines and the Food Pyramid.

Six of the eleven members who were assigned to the U.S. Dietary Guidelines Advisory Committee of the USDA were demonstrated to have financial ties to dairy, meat, and egg industry interests. Prior to the suit, the USDA had refused to disclose such conflicts of interest to the general public. PCRM won the suit in December of 2000.

Dairy milk is being singled out as the biggest dietary cause of osteoporosis, because more than any other food, it depletes the finite reserve of bone-making cells in the body.

Why is this? Although, cow milk contains calcium, much of it is not bioavailable for the human body. Further, animal protein contains a high concentration of sulphur-based amino acids, and consuming the sulphur-based protein in milk makes the body acidic.

Animal protein acidifies the blood, which forces the body to deplete the calcium stored in the bones in order to neutralize the blood.

Acidifying the body's pH causes an immediate biological reaction or correction. Calcium is an excellent acid neutralizer and the biggest storage of calcium in the body is in the bones. The body leaches calcium from the bones in order to balance out the acidity in the blood, which is then expelled through the kidneys and results in a net loss of calcium, actually causing osteoporosis. That's right, drinking milk does not protect the bones from osteoporosis; the opposite is actually the case.

Physicians generally recommend adults take 1000 milligrams to 1200 milligrams. of calcium a day, usually in the form of a supplement. What is the basis for this dietary regimen? The Daily Value (DV), formerly the RDA, is regulated by the FDA and this alphabet soup of regulators arrive at these numbers by making adjustments for the Standard American Diet (SAD), which is based to a large degree on animal product consumption.

They base their calculations on the assumption that the majority is consuming sulphur-based, calcium-depleting protein. The fact is the amount of calcium a person consumes is less important as the amount the body retains. The protein in animal products including beef, fowl, fish, milk, yogurt, and cheese products actually leaches calcium from the bones.[7]

There has been shown to be a strong positive relationship between animal protein consumption and fracture rates. These findings are supported by clinical studies demonstrating that high protein intake worsens calcium losses. A 1994 report in the *American Journal of Clinical Nutrition* illustrated that eliminating animal proteins from the diet cut calcium losses in half.[8]

Dairy products are touted as the antidote to osteoporosis. While they contain calcium, the lactose, naturally occurring growth hormone, antibiotics and other contaminants, cholesterol, saturated fat, and sulphur-based animal protein more than cancel out any possible benefit.

Cow Milk vs. Soymilk

100 grams cow milk	100 grams soymilk
60 calories	43 calories
3 grams of fat	1 gram of fat
10 milligrams of cholesterol	zero cholesterol
full range of amino acids	full range of amino acids
3 grams protein	3 grams protein[9]

Both cow milk and soymilk contain similar amounts of protein and all nine essential amino acids necessary for sustaining life.

Researchers investigating the incidence of osteoporosis in various populations look into the number of hip fractures among elderly women.

The evidence that calcium in milk does not protect against the incidence of osteoporosis has been replicated in numerous studies. For example, in a twelve-year Harvard Nurses' Health Study of seventy-eight thousand women, those who drank milk three times a day actually broke more bones than women who rarely drank milk.[10]

A 1994 study in Sydney, Australia, showed that higher dairy product consumption was associated with increased fracture risk; those with the highest dairy consumption had double the risk of hip fracture compared to those with the lowest consumption.[11]

The incidence of female hip fracture is higher in industrialized countries than in non-industrialized countries and the elevated metabolic acid production associated with a high animal protein diet and ensuing chronic bone buffering and bone dissolution is thought to be a factor. A study published in the journal *Calcified Tissue International* examined the cross-cultural variations in animal protein consumption and hip fracture incidence.[12]

The researchers noted age-adjusted female hip fracture to be higher in industrialized countries than in non-industrialized countries. They hypothesized a possible explanation that had not previously received much attention: that elevated metabolic acid production associated with a high animal protein diet might lead to chronic bone buffering and bone dissolution.

These scientists examined cross-cultural variations in animal protein consumption and hip fracture incidence. Female fracture rates from thirty-four published studies in sixteen countries were analyzed against estimates of dietary animal protein and a strong positive association was found.

The researchers stated that this association could not plausibly be explained by either dietary calcium or total caloric intake and concurred with other studies that suggest the animal protein-hip fracture association could have a biologically tenable basis.

This is not a news flash, as the above-referenced studies were conducted in the 1990s.

It is clear that a diet high in animal protein causes osteoporosis and that women who get their protein from plant foods, such as soy instead of animals, are much less likely to get osteoporosis.

CHAPTER 12

Rumor: Asians Don't Eat Much Soy

Anti-soy claim:
"In fact, the people of China, Japan, and other countries in Asia eat very little soy."—Kaala T. Daniel[1]

There has been much confusion regarding the Asian diet generated by a good deal of misinformation printed and referenced on the Internet.

When confronted by the facts concerning the superior health and longevity of Asian populations who consume the traditional plant-based diet, rich in soyfoods, soy antagonists allege that Asians do not actually consume soy to any meaningful degree.

According to the soy detractors, Asians consume only ten grams of soy per day, which would amount to about two teaspoons. If that sounds preposterous, it is because it is. I would challenge anyone to name a food that is consumed in such small quantities.

Yogurt? Milk? Tea? Ketchup? Broccoli? Carrots? Ice cream? Soy sauce? Of course not; most people take in more salt than that in a day. Because of this contention the amount of soy recommended or even considered reasonable has become a subject of debate.

There are three macronutrients, protein, carbohydrates, and fat and it is disingenuous to portray any food item by one of its nutrients and by doing so the anti-soy campaigners have caused confusion, particularly among consumers.

For example, it has been stated that Asians really consume less than a tablespoon of soy per day. A statement like that would lead the reader to believe that they just crumble two teaspoons of tofu over their rice.

Scientific studies often examine isolated components of various plants, such as protein, antioxidants, or flavonoids. If the claim stated "soy protein," it would be a little closer to the truth. Food is largely comprised of water and it is generally accepted that a tablespoon weighs about fourteen grams.

There are about twelve grams of soy protein in a four-ounce serving tofu and about eight grams of soy protein in an eight-ounce glass of soymilk. Tempeh, a staple soyfood in Indonesia, contains almost sixteen grams of soy protein in a four-ounce serving. That's just one-half cup.

Bill Shurtleff, best-selling author of *The Book of Tofu*, among other soy titles, lived in Japan and traveled widely in Asia studying soyfoods from January 1971 to June 1978.

Shurtleff is the founder of the Soyinfo Center, which houses the SoyaScan Database, the most comprehensive computerized database on soybeans and soyfoods in the world today.

I asked Bill about the proliferation of anti-soy claims in recent years and, based on his personal experiences living among Asians, what he thought of the allegation that Asians do not eat much soy.

Bill had this to say:

> I've seen so many things that they've said that are just so far off, it's ridiculous! Sally Fallon said, in her article in the Weston Price newsletter in 1983, that soyfoods first started to be consumed in Japan after the American Soyfoods association convinced them to do it. They just make hundreds and hundreds of ridiculous statements.
>
> If you presented their case to a typical Japanese, or Chinese, or Korean, or Indonesian educated person, they would look at you and think you were nuts! In the same way that if somebody came to America and they said, you know bread is really awful for you. . . do you realize all of the things it has? You know. . . nobody should be eating bread. People would say, "Where are you coming from?"
>
> They say that people throughout Asia don't eat many soybeans. The Chinese now import 60 percent of all the soybeans produced in the world and the Japanese have soy basically at every meal. Do they expect

> people to have twelve ounces of tofu like Americans have twelve ounces of steak?

Miso soup is among the most well known of Japanese dishes. The Japanese have Miso soup every day, beginning with a bowl at breakfast. In Japan they start the day miso soup like Americans do with their coffee.

Referencing the WAPF, Shurtleff says, "I've seen the statements over and over from them, but all I can say about soy is that there were thirty thousand tofu shops in Japan when I was there. And each of those shops makes their living selling tofu and they make hundreds of pounds a day."

And what about the fertility issue? Shurtleff says:

> If soy had a negative effect on fertility, then the Chinese would certainly have enlisted it in their efforts to control the population.
>
> It's too bad when people get caught up in the little details of things. . . . If you look at the big picture it's obvious that soy does not have a problem with fertility.
>
> People are missing the big picture because of all of the discussion of the added health benefits of soy. There is tremendous importance placed on the soybean in countries that have the best and the greatest longevity in the world. The Okinawans have the greatest longevity in the world. Soy is a very important part of their diet.

John Robbins says it best, "If soy is so bad for you, then how come it is such an important part of the diet of the people who live the longest of anybody in the world?"

Prompted by considerable interest in the possible anti-cancer properties of compounds in soy, Dr. Mark Messina conducted a study regarding the levels of soy consumption in four Asian countries. "Estimated Asian adult soy protein and isoflavone intakes" was published in the journal *Nutrition and Cancer* in 2006.[2]

Among the various suspected chemopreventive compounds in soy, Dr. Messina and coauthors, Chisado Nagata and Anna Wu, chose to look at isoflavones, which have received the most notice.

The team analyzed soy protein and isoflavone intake in the major soyfood consuming countries using individual dietary surveys for the most part.

The study found the highest intake of soy to be in Japan. The results showed that among older Japanese adults consumption of soyfoods averaged six to eleven grams of soy protein per day and from twenty-five to fifty milligrams of isoflavones in the same time period.

Soyfood intake was lower in Hong Kong and Singapore and there were significant regional soy intake differences for China. The authors report that about 10 percent of the Asian population consumes as much as twenty-five grams of soy protein or one hundred milligrams of isoflavones per day.

As a four-ounce serving of tofu contains ten to twelve grams of soy protein and there are three and a half milligrams of isoflavones associated with each gram of soy protein, the same four-ounce serving of tofu would contain between thirty-five and forty-two milligrams of isoflavones. The study would suggest that most Asians eat varying amounts of soy every day.

Best-selling author John Robbins addressed this issue in a letter to the editors of *Mothering Magazine*, referencing an article published in the magazine entitled "The Whole Soy Story," written by Kaayla Daniel, the author of a similarly titled book.

In his words:

> The article's author, Kaayla Daniel, repeatedly says that people of China, Japan, and other countries in Asia eat very little soy, so there is no historical prec-

> edent for eating the amounts being recommended by people like Dr. Andrew Weil and Dr. Christiane Northrup. This is a misleading half-truth.
>
> What's important is not the average soy consumption for the whole of Asia, but the soy consumption in those parts of Asia, which demonstrate the highest levels of human health. And there is no question about where that is.
>
> The elder population of Okinawa (a prefecture of Japan) has the best health and greatest longevity on the planet. . . . The people of Okinawa have repeatedly been shown to be the healthiest and longest-lived people in the world.
>
> This has been demonstrated conclusively by the renowned Okinawa Centenarian Study (OCS), a twenty-five-year study sponsored by the Japanese Ministry of Health.[3]

Elderly Okinawans mortality rates from the multitude of chronic diseases of aging so prominent in the west are among the lowest in the world. As a result they enjoy not only what may be the world's longest life expectancy; Okinawans also have the distinction of having the world's longest health expectancy.

According to the OCS, these Okinawan centenarians have consumed, on average, two servings of soyfoods a day throughout their lives. The OCS examined over nine hundred centenarians and numerous Okinawans in their seventies, eighties, and nineties.

It was clear to the researchers that the Okinawan lifestyle provides many reasons why the elders are so remarkably healthy so far into their senior years. The authors of the Okinawan Centenarian Study state, "Okinawan elders eat an average of two servings of flavonoid-rich soy products per day."

John Robbins continued:

> This is about twenty times more than the amount of soy Kaayla Daniel claims, "Asians really eat." When she says, "there is no historical precedent for eating the large amounts of soyfood now being consumed," she is incorrect. Soy makes up 12 percent of the diet of Okinawan elders. . . .
>
> It is not an accident that in Okinawa, home to the highest soy consumption in the world, heart disease is minimal, breast cancer is so rare that screening mammography is not needed, and most aging men have never heard of prostate cancer.
>
> The three leading killers in the West—coronary heart disease, stroke, and cancer—occur in Okinawans with the lowest frequency in the world.
>
> The authors of the OCS analyzed Okinawan elders' diet and health profiles, comparing them to other elder populations throughout the world. They concluded that the exceedingly low risk for colon cancer and sex organ cancers such as cancers of the breast, prostate, and ovaries credited the high soy consumption in the population.
>
> When compared to Americans, Okinawans have less than half the rate of ovarian and colon cancer and an 80 percent reduced risk of breast and prostate cancer as well. Most would find that to be pretty astounding.

In speaking directly to Kaayla Daniel's assertions, Dr. Mark Messina writes:

> The article "The Whole Soy Story" was filled with so many inaccuracies about soyfoods that it is possible to reach only one conclusion: the author was intentionally misleading readers. Dr. Daniel erroneously stated

> that Asians consume 9 to 36 grams of total soyfoods, or ⅓ to 1½ ounces of these foods per day…
>
> Dr. Daniel confused foods measured in their natural whole food weight (wet weight) with those in the dehydrated (dry weight) state. Readers can understand this concept by simply imagining a pound of apples versus a pound of dried apples; these are two very different amounts of food.
>
> A failure to make this distinction produces gross measuring inaccuracies. In this case, Dr. Daniel failed to understand that 18 grams of dehydrated soy translates into approximately 100 grams of actual soyfoods, or about one serving. One serving provides approximately 7 to 10 grams of soy protein and 20 to 30 milligrams of isoflavones.

According to Dr. Messina, Kaayla Daniel could have readily cited the Food and Agriculture Organization's (FAO) estimates of Asian soy consumption. According to the FAO, Japanese people consume 8.6 grams of soy protein, which translates into about one serving per day.

"Dr. Messina also points out that the FAO averages soy intake from among the entire population. That would have to include infants and children along with adults, therefore that would be presume that the soy intake among the adults would actually be even higher."

Dr. Messina explains:

> Indeed, more precise data coming from at least fifteen large surveys show this to be the case; these surveys have been conducted over the past seven years by scientists in Japan and other Asian countries. This information was readily available to Dr. Daniel since it was published in English-language scientific journals.

> The surveys reveal that Japanese adults consume on average a total of one and a half to two servings of soyfoods per day, about forty to fifty milligrams of isoflavones. Researchers at prestigious Asian institutions such as the Gifu University School of Medicine in Japan have played a key role in providing this information on soy intake.[4]

Dr. Daniel may very well influence some to avoid soy and is very unfortunate for those individuals who may now not enjoy the health benefits demonstrated by Asian populations such as the Okinawans.

The levels discussed here are not only safe they also provide a healthier source of high quality protein, fiber, and health-supporting sub-nutrients and important plant-based variety.

Because of its unique nutritional components, consuming soy in place of the fatty, cholesterol-loaded animal products so prevalent in the American diet has many health benefits.

Six Reasons Why Soymilk Trumps Cow Milk

1. Cow milk provides more than nine times as much saturated fat as soy beverages, therefore cow milk is far more likely to contribute to coronary heart disease.
2. Soy beverages provide more than ten times as much essential fatty acids as cow milk, and therefore soymilk provides a much higher level and far healthier source of these essential fatty acids.
3. Soy beverages are cholesterol-free, while cow milk contains thirty-four milligrams of cholesterol in an eight-ounce cup, which means that consuming cow milk is not the best choice for your heart and cardiovascular system.
4. Soy beverages lower both total and LDL ("bad") cholesterol levels in the bloodstream, while consuming

cow milk actually has the opposite effect: raising both total and LDL cholesterol.

5. Soy beverages contain numerous protective phytochemicals that may protect against chronic diseases such as heart disease and osteoporosis. Cow milk contains no phytochemicals.
6. Men who consume one to two servings of soymilk per day are 70 percent less likely to develop prostate cancer than men who don't.[5]

CHAPTER 13

Soy Bashers: Who Are They Really Hurting?

Much of what we hear from the soy bashing crowd is painted with broad strokes on a canvas crowded with outrageous and deceptive statements that could be characterized as theoretical at best.

As we have learned, these claims are neither supported nor confirmed by any reputable peer reviewed journal or respected medical or nutrition professional in modern health care.

The soy detractors have sought to take down an entire industry with scare tactics and a pattern of relentless repetition that does more damage to health conscious consumers and small entrepreneurs than to large corporations.

The anti-soy contingent characterize themselves as "back to nature" interest groups going up against the forces of "corporate evildoers." It is oftentimes easy to get behind a cause or concept that seems to take on big business as we all struggle to maintain some control in our own lives; however, soybean farmers in America come in all sizes.

Jonathan Chambers is a fifth generation family farmer who has been growing soybeans in Iowa for twenty-five years. Chambers Family Farms market their Laura® Soybeans internationally. Laura® Soybeans (named for Jonathan's sister, Laura Jo) is a strain of soy developed by the Chambers family over generations, the old-fashioned way. They have employed careful, non-GMO breeding over years of sustainable farming practices, which would make Mendel proud.[1]

Laura® soybeans are premium soybeans prized for their sweet flavor and are considered the best soybeans for making soymilk and tofu by just about everyone in the know, myself included. The Japanese have been importing Laura® soybeans for many years for just that purpose.

Growing soybeans is the family business, and the volume and intensity of the anti-soy claims has been frustrating for operations like the Chambers Family Farms, who depend on health conscious consumers to support their enterprise.

Many consumers have been confused by the frightening allegations and do not know what to think, often erring on the side of caution and avoiding soy entirely.

In the final decades of the twentieth century, almost five million family farms were replaced by large-scale manufacturing operations. A small family farm by some standards, the Chambers Family has seen some pretty dramatic changes in their rural farming community in north-central Iowa.

Modern agriculture and industrialized animal production has had a dramatic impact on society, human health, and the environment. According to Henning Steinfeld, a senior U.N. Food and Agriculture Organization (FAO) official, "Livestock are one of the most significant contributors to today's most serious environmental problems."

Steinfeld is senior author of *Livestock's Long Shadow*, an in-depth FAO report meant to draw attention to the very substantial contribution of animal agriculture to climate change and air pollution; to land, soil, and water degradation; and to the reduction of biodiversity. Steinfeld added, "Urgent action is required to remedy the situation."[2]

The U.N. Food and Agriculture study also reported that "farm animals" generate more greenhouse gas emissions than cars and place an enormous burden on increasingly scarce water resources. Livestock now use 30 percent of the earth's entire land surface and pose one of the greatest threats to climate, forests, and wildlife.

Mostly permanent pasture, 33 percent of the earth's arable land is used to produce feed for livestock. Forests are cleared to create new pastures and this is the driving force behind deforestation, especially in Latin America where some 70 percent of the Amazon forests have been turned over to grazing.

Using soybeans to make milk is a viable alternative to the raising of dairy cows with definite ecological advantages. Certainly the sheer volume of the soybean crops that could be grown on the same land would feed people many times over than if it were used to raise cows.

To be clear, the overwhelming majority of the soy raised in this county and around the world is produced for animal feed. It is grown to feed cows, pigs, chickens, and turkeys. Soybeans are fed to huge herds of livestock who foul the earth, water, and air and who are eventually slaughtered and fed to people. Old dairy cows, no longer able to produce milk, become the source of cheap product for the meat industry.

Many scientists link the gases that trap heat in the atmosphere, often called greenhouse gases, to global warming.[3] [4] An overwhelming body of evidence associates a growing global appetite for animal products with the looming health and obesity pandemic and escalating destruction of our ecosystem.[5]

Research published more than a decade ago in the *Journal of Clinical Nutrition* used nutrition ecology, an interdisciplinary science that encompasses the entire nutrition system, to show how vegetarian diets protect the environment, reduce pollution, and slow global warming.[6]

Animal husbandry is a major factor in water pollution, contributing animal waste, antibiotics and hormones, and a witch's brew of chemicals from tanneries, fertilizers, and pesticides from feed crops.[7]

Raising cows, whether for meat or dairy production requires far more energy. In terms of animal feed, one cow consumes about fifty-three pounds of dry food and from twenty-five to fifty gallons of water every day. A dairy cow will produce an average of forty kilograms or eleven gallons of milk a day.

Soybeans can produce at least twice as much protein per acre than any other major vegetable or grain crop and up to fifteen times more protein per acre than land used for meat production.

Soybeans are the number two US export, with corn being the number one crop exported in America, and according to Jonathan Chambers, corn and soy is the crop rotation. If you are growing soy, you are growing corn.

"That is what replenishes the soil, the soybean and corn crops work together as the soy 'sticks' the nitrogen for the corn. For example, if you get forty bushels of beans then you can say you are going to have forty pounds of nitrogen per acre for the corn the next year."

Legumes are particularly important for their ability to fix atmospheric nitrogen naturally, a process that reduces the cost of fertilizer for farmers. This feature is what makes legumes the best choice for crop rotation. Soybeans replenish the valuable nitrogen in the soil in which it is grown because they have nodules on their roots that contain nitrogen-fixing bacteria.

Soy had a very important role in the America following the First World War. The Great Depression, which began in 1929, was a period of severe worldwide economic depression and the worst economic depression in US history.

From 1930 to as late as 1940, severe dust storms caused major agricultural and ecological damage to the American and Canadian prairie, which came to be called the "Dust Bowl."

This devastating phenomenon was caused by severe drought and decades of widespread farming without crop rotation, cover crops, or other techniques that would have prevented wind erosion. Soybeans came to the rescue in the drought stricken Dust Bowl regions of America, where farmers were able to use soybean crops to regenerate the soil with its nitrogen-fixing properties.

I asked Jonathan about the impact of factory farming on Corwith, Iowa, the farming community that includes the Chambers farm.

> Last year a company came in here and built a huge chicken egg–laying facility southeast of our town with many millions of chickens. The stench is horrendous; it absolutely reeks.
>
> Over the last twenty-five years farming has changed drastically. First of all, back then we had a lot more neighbors. There were more families in the community. There were more farms between four hundred and eight hundred acres per family. There were more kids in the schools and no vacant houses in our town. There were more small businesses in our community.

> Young people are not coming back and now we have a lot of transients.

Is he optimistic?

> I try to be. My son wants to come back and I expect he will. Dad and I are going to do everything we can to make that possible. And our neighbor's son wants to come back. Right now, all of a sudden there seems to be a renewed interest in young people wanting to come back, which is great.
>
> Unless someone is well-capitalized from other ventures or maybe has won the lottery, it would be impossible for most people to start farming on their own. So, it would have to be passed from one generation to the next.
>
> The farms have gone from four hundred to eight hundred acres to farms that are one thousand, two thousand, five thousand acres to what I call, "Wildcat Farmers" who are farming ten thousand to twenty thousand acres. They don't actually ever touch the soil; they are more businessmen who hire people to do all the equipment.
>
> We have had no increase in the student population at our local school because this company is here. I haven't seen any new people in the community and the only thing they have done in town has been to take a small building and renovate it into an office with a few cars out in front. Hardly what anyone would call a boon to the local economy.

According to the National Soybean Research Laboratory, soybeans can produce at least twice as much protein per acre than any other major vegetable or grain crop, five to ten times more protein per acre than land set aside for grazing animals to make milk, and up to fifteen times more protein per acre than land set aside for meat production.

Perhaps the most distressing aspect of this entire soy-bashing episode has been the impact of all the hype on those who would benefit the most. The epidemic of obesity and its life threatening complications is directly related to the Western diet, loaded with saturated fat and cholesterol,[8] which is only found in animal products.[9]

Even if soy did not have any of the unique health benefits attributed to it, the high level of complete protein makes soy a natural replacement for animal products. Almost 40 percent of the calories in soybeans come from protein and that means that soybeans are higher in protein than other legumes and even many animal products.

According to the World Bank, malnutrition remains the world's most serious problem and the single biggest factor to child mortality. Soyfoods are an excellent low-cost protein solution for feeding the world's hungry.

The Global Health Council (formerly the National Council of International Health) is a nonprofit organization that was created in 1972 to identify world health problems and report on them to the US public, legislators, international and domestic government agencies, academic institutions, and the global health community.

Based in Washington, DC, it is the world's largest membership alliance dedicated to saving lives by improving health throughout the world. Members include healthcare professionals, government agencies and non-government organizations, foundations, corporations, and academic institutions.

UNICEF is an acronym that stands for the United Nations International Children's Emergency Fund, and it was established in 1946 to meet the emergency needs of children in post-war Europe and China. In the early 1950s, UNICEF became a permanent part of the United Nations and its mandate was broadened to address the long-term needs of children and women in developing countries everywhere.

UNICEF defines undernutrition as the outcome of insufficient food intake and repeated infectious diseases. It includes being underweight for one's age, too short for one's age (stunted), dangerously thin for one's height (wasted), and deficient in vitamins and minerals (micronutrient malnutrition).

According to the Global Health Council, undernutrition, too few nutrients to sustain healthy normal growth and development, contributes to 3.5 million maternal and childhood deaths in developing countries around the world annually.

Soybeans Can Feed the World

The Food and Agriculture Organization of the United Nations (FAO) report "The State of Food Insecurity in the World 2015" estimates that there are 793 million hungry people in the world. The FAO defines hunger as being synonymous with chronic undernourishment.

It is essential to provide a sustainable source for high-quality protein and the critical nutrients necessary for survival. The soybean contains complete high-quality protein and unique properties that make soy the best choice for alleviating hunger.

Soyfoods can help end the issue of hunger, which is faced by millions in developing countries. Farmers are able to grow more soybean crops with fewer natural resources than any other good source of protein because soybeans require less water and less energy to cultivate and they take up less land.

The World Soy Foundation (WSF) is the philanthropic arm of the American Soybean Association, formed in 2006. The WSF works with impoverished communities teaching business skills and the importance of good nutrition and provides disaster relief.

The WSF is dedicated to the relief of hunger and malnutrition in the world by helping to coordinate programs that recognize the importance of the use of soybeans in developing sustainable food solutions. They focus on building microenterprises and small businesses utilizing soy, from production to consumption.

SoyCow Machine

The WSF's mission is to reduce malnutrition through the power of soy, and to that end they develop sustainable solutions through soybean-based nutrition. Good nutrition is the first line of defense against childhood disease and is essential in order to improve children's potential to build healthy, peaceful, and thriving communities.

The WSF has developed a number of innovative programs to meet the needs of disadvantaged populations in the developing world, such as the SoyCow and the VitaGoat.

The SoyCow is a machine that processes whole soybeans into soymilk from start to finish. The soymilk not only provides wholesome beverages, it can also be the first step in the making of a number of other soyfoods.

What sets the SoyCow apart from the automatic soymilk makers many consumers have in their own kitchens is the volume it can produce. The SoyCow can process four pounds of raw soybeans into four gallons of nutritious soymilk in about twenty minutes.

This soymilk can then be further processed into highly nutritious soyfoods, such as tofu and yogurt as well as local food blends. A byproduct of soymilk making is the production of copious amounts of the protein and fiber-rich pulp okara. Okara has myriad uses and can be used to make okara burgers and loaves and to enrich baked goods, spreads, and dips.

SoyCows are particularly well suited for developing countries with low labor costs. The operation of one unit can create employment for up to six unskilled people while providing critical nutrition to hundreds of people.

The VitaGoat is a food processing system that can make many value-added products from various cereals, grains, nuts, fruits, and vegetables. This innovative system can enable local people to increase their food security, improve their health, and create micro-businesses and employment. The VitaGoat can process primary foods into flours, pastes, or wet slurries such as soya mash, which can be used as is or cooked with steam.

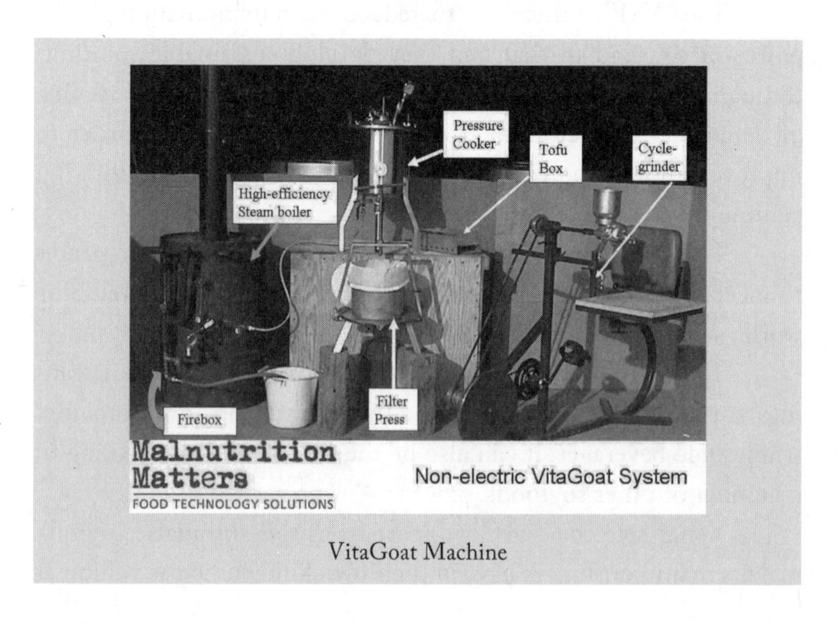

VitaGoat Machine

The significant feature of the VitaGoat machine is that it can make all of these foods without electricity; grinding is provided through "pedal power" while cooking energy is provided by the innovative and fuel-efficient steam boiler.

In recent years VitaGoat pedal-powered machines have been installed in thirteen African countries, India, Thailand, Bangladesh, and North Korea. In Orissa, India, a women's self-help group uses the VitaGoat for their mid-day meal program. WSF also runs training and support centers and is developing an innovative program, the Youth We Feed Can Lead, in Guatemala. This pilot program combines nutrition education and video production training and equipment to empower youth and the community to be agents of change.

Abbie Mchunu is a member of the Inkatha Party of the South African National Government. She is known as "Mama Soya" for her longtime and personal support of expanded use of soy in Southern Africa, and she once drove for more than three hours by herself in a pickup truck to get to a soyfoods demo.

Ratan Sharma cutting freshly pressed tofu for participants of a training session demonstration in soyfoods production. Onlookers include Henry Davies (right) and Abbie Mchunu (center in blue).

CHAPTER 14

Scientific Research Defined and Explained

Medical science is always evolving and tends to move quickly. Scientific journals are periodicals that report on new research and the publication of the results of research is an essential part of the scientific process. These journals are highly specialized and provide detailed analysis allowing researchers to keep up to date with developments in their field and direct their own research.

As of October of 2015, there were more than ten thousand open access journals, which represent just 12 percent of journals. Now in the age of the Internet, if you know what to look for, research findings are just one or two clicks away.

Throughout this book we reference research that has been "peer reviewed," which means that the article was reviewed and critiqued by the author's peers, who are experts in the same subject area. Peer-reviewed articles have been evaluated by a board of impartial reviewers for quality of research and compliance to editorial standards before publication.

The myriad of anti-soy claims tend to cite one scientific study or another hoping to lend an air of legitimacy for the general reader, because most people don't generally understand how empirical science is done.

The references cited by the anti-soy critics like the WAPF are taken from animal studies. They cite studies where subjects were injected with high volume of elements extracted from the soybean with adverse effects while ignoring research where human subjects consumed actual soyfoods and experienced improved health.

In order to investigate these rumors with a degree of insight, it is helpful to gain an understanding of the types of methods and models that are used in the scientific process of experimentation.

When reviewing the scientific literature, it is important to note the intended goal of the researchers. What is the object of the study and what is the hypothesis (supposition that provides the starting point for further investigation)? What is the nature of the experiment and does it apply to human physiology? The goal and the methodology are key when applying the results of a particular study, as are the applications, benefits, and limitations. Like most things, once you have a working knowledge of the process, it is not as complex as you might have thought.

The origin of the word "research" is from the French *recherch-er* (ray- cher–shay), which means to search closely. Research science is based on the application of particular scientific methods in order to provide theories to explain the nature and properties of the world around us.

Science is an ongoing process of discovery. Scientific research is defined as the performance of a methodical study, which follows a series of steps and strict protocol in order to answer a specific question or prove a hypothesis. Scientists are trained to use the scientific method, which is a step-by-step process for reaching a logical conclusion to any question.

Scientific research requires critical thinking. Scientific study and experimentation use logical reasoning, which is the strategy that scientists apply in order to understand how everything works.

All scientific studies begin with a research problem framed as a question that leads to a hypothesis. The hypothesis is a tentative explanation for a particular occurrence that is assumed to be true for the sake of argument. The ultimate goal of any experimental process is to find a definitive answer. It should be noted here that research studies often conclude with more questions than answers, as unforeseen issues often become apparent in the course of the study.

Research science is conducted using the accepted scientific method, which is a four-step process that begins with a description of what is to be observed, the formation of the hypothesis, a prediction of the outcome, and performance of the experiments and corresponding tests.

The hypothesis is the basis of the study and is used to make the predictions, which can then be tested by observation of the outcome. If the outcome of the experiment is inconsistent with the hypothesis, then the hypothesis is rejected. However, if the outcome is consistent with the hypothesis, the experiment is then said to support the hypothesis. These are the basic rules, which can vary slightly among different fields of science and may be scientific, historical or economic.

There are two fundamental types of experiments: in vitro (in ve'tro), which literally means "in glass," and in vivo (in ve'vo), which is Latin for "within the living." Tests performed in vitro are those that are done in vessels, such as test tubes or in a petri dish. The petri dish contains a seaweed-based agar jelly where bacterial cultures are studied in the laboratory.

In vivo experiments are done using a living organism. Animal testing and clinical trials are two kinds of in vivo research. In vivo testing is often favored over in vitro because it affords the opportunity to observe the overall effects of the experiment on a living subject. This type of experimentation is often described by the axiom "in vivo veritas."

In vivo studies are divided into two classifications: animal testing and clinical trials. Animal testing or experimentation employs the use of nonhuman animals for research. The number of vertebrate animals, such as primates, used in medical research annually is estimated to be in the tens of millions and may be as many as one hundred million, worldwide.[1] The invertebrates such as mice and rats, birds, fish, frogs, and animals who are not yet weaned are not counted in the possible one hundred million lab species and one estimate places the number of mice and rats used in the US alone at eighty million.[2]

Animal Testing

Animals have been used throughout the history of biomedical research with the earliest references in the writings of the ancient Greeks in the fourth century BC. A number of scientists and animal welfare and animal rights organizations such as Physicians Committee for Responsible Medicine (PCRM),[3] People for the Ethical Treatment of Animals (PETA), and Cruelty Free International question the legitimacy of animal experimentation, arguing that it is cruel and in fact poor scientific practice.

These groups are joined by a growing number of organizations such as the Humane Society of the United States (HSUS), who contend that testing on animals is poorly regulated and that medical progress is actually being hindered by misleading animal models, that some of the tests are outdated, that it cannot reliably predict effects in humans, that the costs outweigh the benefits, and that animals have an inherent right not to be used for experimentation.

Clinical Research

The National Institutes of Health (NIH) defines clinical research as patient-oriented research conducted with human subjects where an investigator directly interacts with human subjects.

Clinical trials conducted with human participants determine whether biomedical or behavioral interventions are safe and effective. The clinical trial is carried out so that data can be collected for behavioral or health interventions that can include diet, therapy procedures, pharmaceutical drugs, diagnostics, physical activity, or medical devices.

Investigators initially enlist healthy volunteers and patients into small-scale pilot studies, followed by larger studies. Clinical trials may vary in size from a single center in one country to multi-center trials conducted in multiple countries.

In the clinical trial, investigators recruit patients with predetermined characteristics. They then administer treatments or behavioral interventions, while collecting ongoing data that documents the patients' health for a specified period of time. The data includes the measurement of vital signs, the concentration of the substance or drug being studied, and whether or not the patient's health improves. The researchers then have the pooled data statistically analyzed.

Types of Clinical Trials

Treatment Study

Randomized controlled trials (RCT) are the most common type of scientific experiment among treatment studies. The distinguishing feature of the RCT design is that the participants are randomly chosen to receive one or another of the treatments under study. After randomization, the groups of subjects (two or more) are followed up in exactly the same manner. RCTs are sometimes known as randomized controlled clinical trials.[4] An important advantage of randomization is that it minimizes bias and balances out the known and unknown factors in the process of assigning the mode of treatment.[5]

Blinded Trial

Blinded is used figuratively here, as in blindfolded. In this sense some of the individuals involved are prevented from certain knowledge that may affect the outcome. RCT's may be blind studies where some of the subjects of the study are prevented from being aware of any information that could lead to bias, conscious or unconscious, on their part that would invalidate the results. In the single-blind study information that could cause bias or otherwise skew the results is withheld from the participants; however, the experimenter is in full possession of the facts.

A classic example of a single-blind test is the taste test often employed in marketing. Several unbranded samples are labeled A, B, and C and volunteers are asked to sample the product. The marketing person knows which sample is the branded product, but the participants do not. The problem with the single-blind test is that the person giving the test could give cues that may well be subconscious, creating bias and skewing the results.

Double-Blind Trial

In the double-blind trial neither the individuals nor the researchers are aware of key information, such as which individuals are in the control group and which are in the experimental group. The double-blind method is a more strictly controlled method of conducting an experiment with regard to eliminating subjective bias on the part of all the parties involved.

Random assignment of the subject to either the experimental group or the control group is a critical aspect of the double-blind research design. The identity of the subjects and which group they are a part of is kept by a third party and not given to the researchers until the study is completed. Double-blind experiments lessen the influence of conscious or subconscious prejudices and inadvertent physical cues, which can skew the results.

Epidemiological or Observational Study

Epidemiology studies populations and is most often used to examine patterns of health, illness, and related factors to identify risk in public health research.

In epidemiological studies the researcher observes but does not alter what occurs and can be described as behavioral in that the subjects' behavior is observed throughout the trial.

One type of observational study is the cohort study. A cohort would be a group of people who share a common characteristic or experience within a specific period of time. For instance, they may have been exposed to a particular drug or a vaccine, so that a group of people who were born within a particular time frame, say 1960, would comprise a birth cohort. The comparison group may be the general population from which the cohort is drawn or it may be another cohort of persons thought to have had little or no exposure to the substance under investigation but are otherwise similar.

A prospective cohort study follows a group of similar individuals over time. The "cohort," would differ with regard to specific factors under study. Prospective cohort studies are done in order to determine how certain factors affect the rates of a particular outcome. Prospective studies allow researchers to study of the origin of disease and disorders in humans in controlled experiments without deliberately exposing subjects to suspected risk factors, which would be unethical.

Case Controlled Study

In the case control study, individuals with a specific disease are compared with patients who do not have the disease. Case control studies are also known as retrospective studies, which look back retrospectively to compare how frequently the exposure to a risk factor such as a history of smoking or obesity is present in each group. Those without the disease comprise the control group.

Case studies allow the researchers to look at multiple risk factors at a time and can detect trends or patterns of disease over time and can help to estimate odds. Retrospective studies have some issues with data quality in that they rely on memory. People with the condition will be more motivated to recall risk factors (recall bias).

Cross-Sectional Study

A cross-sectional study or analysis is the simplest type of observational study sometimes carried out to investigatc associations between risk factors and outcome. The cross-sectional study has been described as a snapshot of the incidence and characteristics of a disease in a specific population at a particular point in time.

Cross-sectional studies provide data on the entire population under study, whereas case control studies typically include only individuals with a specific characteristic, along with a small sample of the rest of the population.

Meta-Analysis

Meta-analysis can be described as the process of conducting research about previous research.

Statistics is the study of the collection, analysis, interpretation, presentation, and organization of data. Meta-analysis is a systematic review whereby the findings of a number of previously conducted studies that address a common problem are combined statistically.

Meta-analysis gives researchers the ability to contrast results from different studies. This makes it easier to identify patterns among study results as well as sources of disagreement and other in relationships that may come to light.

The conclusion drawn from meta-analysis is considered to be statistically stronger than the analysis of any single study. This is due to the increased number of subjects, the potential for greater diversity among subjects, and the likelihood of accumulated effects and results.

This type of analysis can provide a summary for a particular area of research and by combining the results of diverse statistical studies meta-analysis increases the number of subjects, giving more weight to the findings.

This summary is a simplified account of some of the methods and procedures used in scientific research. The researchers interviewed for this book have made numerous contributions to evidence-based science and have generously shared their deep knowledge and familiarity regarding the characteristics of soy. It is my intention here to provide some background and a brief explanation of the procedures common to peer-reviewed scientific studies for the reader.

CHAPTER 15

So What Is the Skinny on Soy?

We have examined each of the claims that are most often cited by those who seek to discredit soy and presented prominent researchers in the field. In every case, the evidence debunks the theories that have been so widely circulated for at least the last fifteen years.

When you shine the light of common sense on these rumors, they do not hold up. Probably the most alarming allegation is the claim that eating tofu or drinking soymilk could affect fertility or cause birth defects.

Is it really plausible that consuming soyfoods could have a feminizing effect on young men or cause sterility? To believe that one would only have to look to the East where overpopulation is the issue, not infertility. And now, Internet headlines like "Soy is making our kids gay" have taken this rumor to a whole other level.[1]

This article has been referenced and reprinted all over the Internet, and its author, Jim Rutz, is neither a health professional nor a research scientist. He is a minister.

Here is an excerpt:

> There's a slow poison out there that's severely damaging our children and threatening to tear apart our culture. . . . When you eat or drink a lot of soy stuff, you're also getting substantial quantities of estrogens. . . . Soy is feminizing, and commonly leads to a decrease in the size of the penis, sexual confusion and homosexuality.

Such over-the-top pronouncements would be hilarious if it did not scare the bejesus out of new parents. I have seen websites with numerous comments from overwrought parents of infants who are feeding with soy formula.

Prominent scientists in the field have shared their experiences in previous chapters. Soy does not contain the female sex hormone, estrogen. This is not speculation or a theoretical statement; this is an actual fact. In scientifically validated studies, the preponderance of evidence demonstrates that soy does not lower testosterone levels or raise the level of estrogen in the body.

Soy does contain phytoestrogens, and while the body might be fooled into thinking it is estrogen, it is not estrogen. The fact that the human body accepts phytoestrogens into the cells as though it were estrogen is very beneficial, as the much weaker phytoestrogens replace the actual estrogen in the body, potentially reducing the levels of estrogen dramatically.

Why is it important to reduce the levels of estrogen in the body? Numerous of studies suggest a link between high levels of estrogen and breast cancer. One such study out of Australia investigated a particular gene known to promote the growth of cancer that can be "turned on" by estrogen.[2]

The cancer biology team from the University of Queensland's Diamantina Institute for Cancer, Immunology and Metabolic Medicine believe their research will help explain the link between breast cancer and high levels of estrogen.

The Brisbane scientists' findings demonstrate that estrogen turns on a gene linked to breast cancer. Research leader Professor Tom Gonda said, "What we've shown is that the ability of estrogen to switch this gene on is important for the growth of breast cancer cells."

If soy were problematic because it contains phytoestrogens, then the same logic would apply to flax seeds, which are very high in phytoestrogens (higher in fact than soybeans). Numerous seeds and nuts—particularly walnuts, breads, whole grain cereals, and legumes like garbonzo beans—all contain phytoestrogens. And this is a good thing, because these plant-based estrogens counter the adverse effects human estrogen can exert in the body.

It is quite interesting that those who would create shock waves in the media at the thought of ingesting a substance that purportedly contains estrogen are the same people who recommend consuming copious amounts of dairy products that are loaded with sex hormones.

Cows are massive. They are many times the size of a human and, as such, have a huge amount of alien hormones that occur naturally, not to mention the additional bovine growth hormone rbGH, used to increase milk production since its approval by the FDA in 1993.

What most consumers do not know is that dairy cows are all pregnant and this introduces a range of other hormonal issues, which place the human body at risk for any number of diseases.

According Dr. Ganmaa Davaasambuu, a scientist at the Harvard School of Public Health, "Among the routes of human exposure to estrogens, we are mostly concerned about cow milk, which contains considerable amounts of female sex hormones. Dairy," she added, "accounts for 60 percent to 80 percent of estrogens consumed."

Dr. Davaasambuu, who is also a Mongolia-trained medical doctor, a Japan-trained PhD in environmental health, and a current fellow at the Radcliffe Institute for Advanced Study, told an audience of her fellow Fellows at Radcliffe:

> Part of the problem seems to be milk from modern dairy farms, where cows are milked about three hundred days a year. For much of that time, the cows are pregnant. The later in pregnancy a cow is, the more hormones appear in her milk.
>
> Milk from a cow in the late stage of pregnancy contains up to thirty-three times as much of a signature estrogen compound (estrone sulfate) than milk from a non-pregnant cow.

The typical dairy cow is milked for ten months of the year. This is only possible because dairy cows are impregnated by artificial insemination while still secreting milk from the previous pregnancy. Milk from pregnant cows contains a far higher level of hormones than the milk obtained from cows who are not pregnant. Even on organic dairy farms, dairy cows are kept constantly pregnant.

What about the phytates? Phytates are credited with antioxidant properties, may strengthen the immune system, and may help to prevent and treat cancer. Are these substances found exclusively in soy?

If you only listen to the anti-soy crowd, you might believe that soybeans alone contain phytates.

Not true. Phytates are found in grains, legumes, nuts, and rice—more so in whole grain products than in processed foods as it resides in the bran, which is practically eliminated in processing.

According to *Food Phytates*, a book that looks into the potential health benefits of phytates, the binding properties of phytates have caused them to be viewed traditionally as anti-nutrients. However, research suggests that phytates may actually reduce the risks of cancer and heart disease. The authors provide numerous tables and figures and list the percentage of phytates in soymilk as 0.11 percent and note that Durham wheat contains eight times more phytates than soymilk at 0.88 percent.[3]

Do they inhibit the absorption of minerals? Yes, to a degree and if that is a concern, consuming foods rich in vitamin C with legumes or whole grain foods will increase the absorption of minerals. How does a glass of orange juice with a bean burrito sound? Vitamin A has also been shown to reduce the inhibition of iron absorption,[4] so you might want to add a few raw carrots before you roll up that tortilla.

Common sense dictates and nutritionists agree that a balanced diet of plant foods that includes a varied selection of vegetables, fruits, grains, and legumes is healthful and nourishing. Those who would like us to believe that plant foods somehow contain "poisons" and that flesh foods are healthful and nutritious also posit that dietary cholesterol promotes a healthy heart and saturated fat does not clog arteries.

Can anyone seriously entertain the idea that eating tofu leads to dementia? The facts just do not support such an assumption. We need to turn our attention again to the East where education and discipline are a central theme. Asian students far outshine their American counterparts and while culture may play a role, a diet rich in soyfoods does not seem to have interfered with their brain function.

Soybeans have been consumed by humans in Asia for millennia in the form of tofu, soymilk, tempeh, and natto. Now, soyfoods have become popular in other parts of the world and modern food science has developed soy alternatives that recreate familiar foods in a healthier and more nutritious way.

These convenient, "second generation" soyfoods have broad appeal and consumers can find everything from tofu manicotti to soy cheeses, yogurt, dips, and non-dairy frozen desserts.

Choosing soyfoods over meat and dairy products has long been recommended as the healthiest of food choices. Research continues to demonstrate that a plant-based regimen of vegetables, legumes, grains, and fruit not only supports human health and nutrition, but is optimal.

The anti-soy crowd would have you believe that we should eat like cave men and promote what has been dubbed the Paleo Diet, which is built heavily on flesh foods and excludes all grains and legumes. This is a regimen that recommends consuming as much as 65 percent protein, a prospect that should concern anyone who wants to hold onto his or her kidneys.

The government, media, advertising agencies, and industry public relations firms have all done quite a job convincing the public that eating animals is tasty, healthy, and downright necessary! In the end, however, it has all been about commerce.

Most people would be surprised to learn the Dairy Board is little more then a public relations agency created to promote milk products. It is more accurately described by its formal title, the "National Fluid Milk Processor Promotion Program," which was first authorized by the Fluid Milk Promotion Act of 1990.[5]

John Robbins is a best-selling author and visionary. I have a great deal of respect for his work and the way he has lived his life. In this quote from *The Food Revolution*, John speaks to mainstream perceptions and the way many view healthy plant-based food choices:

> The standard diet of a meat-eater is blood, flesh, veins, muscles, tendons, cow secretions, hen periods and bee vomit. And once a year during a certain holiday in November, meat-eaters use the hollowed out rectum of a dead bird as a pressure cooker for stuffing. And people think vegans are weird because we eat tofu?[6]

The widely circulated claims that consuming soyfoods will result in feminized characteristics and that soybeans contain poisons, cause birth defects, cancer, homosexuality, Alzheimer's Disease, thyroid problems, or bone deficiencies quickly fall apart in the light of real science.

A Word about GMOs

"Genetically modified organism" defines food products that have been altered at the gene level with biotechnology.

The GMO versus non-GMO controversy has been ongoing, with consumers worried about the safety of genetically engineered food and the reluctance of the industry to label such products. The debate regarding the safety of GMO food crops has been a controversial media topic and there have been some troubling animal studies reporting negative health effects.

Consumers and environmental groups have been working toward the adoption of state and federal regulations regarding the labeling of genetically modified crops, as many consumers become more and more concerned about the corporate monopolization of our food supply.

GMO plants are transgenic in that these plants contain one or more genes that have been introduced by artificial methods, which include gene transference and cell fusion in place of traditional breeding.

Genes of unrelated species such as animal, bacterium, or virus introduced into a different organism (corn, soybeans, cotton, etc.) will irreversibly alter the genetic code of the plant.[7]

Modern biotechnologists have created tomatoes with a longer shelf life by adding flounder genes—that's right . . . fish! They have engineered soybeans resistant to weed killers and potatoes with jellyfish genes that glow in the dark when they need water. If you find this disturbing, you are not alone.

Scientists have genetically modified the DNA of major agricultural crops such as soybeans, corn, and cotton in order to resist a widely used herbicide. Roundup is the flagship of Monsanto's agricultural chemicals business and these new genetically modified plant strains are referred to as "Roundup Ready."

It is true the most of the soybean crop is GMO; however, it is important to note that about 70 percent of the soybeans are processed for animal feed. Approximately 87 percent of all soybeans worldwide are crushed into soybean meal and soy oil and the remaining 13 percent is used for direct human consumption.

The annual report on the worldwide commercial use of genetically modified plants is published by the agro-biotechnology agency, the International Service for the Acquisition of Agri-Biotech Applications (ISAAA). According to the 2013 report, the field area of GMO soybeans compared to the total soy production decreased to 79 percent.

DuPont and Monsanto charge twice as much for their GMO seed corn and soy and justify the cost with the potential of greater yields and ease of production. However, many farmers believe that the extra investment in seed and attendant chemicals is not covered by the possible bump in yield.

Because of the consolidation of the seed business, many public varieties of seed produced with natural breeding techniques from breeding programs at institutions such as the Universities in Minnesota, Michigan, Iowa, Nebraska, and Illinois no longer supply the lion's share of product.

If the foods that you purchase are 100 percent organic that is one way to avoid having genetically modified ingredients on your plate. In the US and Canada regulations prohibit companies from labeling products "100 percent certified organic" if they contain genetically modified ingredients.

Although the industry is averse to labeling their GMO products as such, the consumer can discern between GMO and non-GMO by looking for the NON-GMO label. Soy products that are non-GMO proudly display it on the front of their packaging and provide the information on their websites.

In North America more than 70 percent of packaged foods contain GMOs and ingredients such as corn syrup, cornstarch, processed sugar, canola oil, and various food additives and flavorings are common in packaged goods. It pays to be vigilant, and if organic is not always possible, the non-GMO or verified non-GMO label is your best bet.

Buying local at farmer's markets is also a great option. Know your farmer, like I do when buying my non-GMO Laura® Soybeans from Jonathan Chambers, of Chambers Family Farms. We all need to know as much as we can about the source of the food we put into our bodies.

THE SOYFOODS PANTRY

Miso

Miso is a naturally fermented paste made by combining cooked soybeans, salt, and sometimes other ingredients, such as rice or barley. This flavorful paste is used to enhance the flavor of sauces, soups, dips, marinades, dressings, and main dishes. Miso is high in phytochemicals, beneficial enzymes, and bacteria that help keep the gut healthy.

Miso comes in a variety of flavors textures, colors, and aromas, and the flavor profile differs greatly with the color ranging from bright gold to blackish-brown. The longer miso is aged, the deeper in flavor it gets and, generally, the lighter in color the miso the sweeter and less salty it is. Light colored miso is younger than darker ones. Miso is gluten-free, unless it contains barley or wheat.

There are three main categories of miso: rice miso, barley miso, and strictly soybean miso.

Four Popular Miso Varieties:

1. White (shiro) miso is milder and has a sweet, mellow flavor and is lower in salt than other miso. Sometimes called "mellow white miso," it is ac-

tually very pale yellow, and is a great addition to dressings, sauces, and soups.

2. Yellow (shinshu) miso is often made with barley, fermented a little longer than white miso and can be yellow in color to light brown. Yellow miso adds a deeper flavor to soups and glazes.
3. Red (aka) miso is the saltiest and most pungent variety. Red miso is fermented the longest and is darkest in color. Because of its strong flavor, red miso is used more sparingly, adding bold flavor to any dish.
4. Hatcho miso is referred to as the "miso of emperors" and is pure soybean paste made with a special type of koji or culture. Hatcho miso is aged for at least sixteen months and is reddish-brown, chunky, and often used to flavor hearty soups.

Natto

Natto is made from fermented, cooked whole soybeans and has been consumed for centuries in Japan. Natto has a sticky, viscous coating and somewhat cheesy texture and, because the fermentation process breaks down the beans' complex proteins, natto is more easily digested than whole soybeans. In Japan, natto is flavored with soy sauce and mustard and used as a topping for rice, or added to miso soup or vegetables. Natto can be found in Asian and natural food stores.

Nondairy Soy Frozen Dessert

Soy-based nondairy frozen desserts are made from soymilk or soy yogurt. Soy ice cream is one of the most popular desserts made from soybeans and can be found in natural food stores and many supermarkets.

Okara

Okara is the pulp fiber that is the byproduct of making soymilk that is rich in protein and fiber. When dried okara has a texture similar to shredded coconut that adds protein and fiber to baked goods. Okara makes delicious burgers and adds bulk and nutrition to cookies and quick breads.

Anyone who makes homemade soymilk has an abundant supply of okara on hand. If you are not someone with a nifty automatic soymilk you can find okara in Asian groceries and online.

Soybeans, Green Vegetable, Edamame

Edamame (pronounced ed-dah-MAH-may) or green vegetable soybeans are immature soybeans harvested at 80 percent maturity. They are larger and more oval in shape than either yellow or black soybeans. Edamame is available in the pod or shelled, frozen or fresh, in natural food stores, Asian markets, and in an increasing number of traditional supermarkets. Edamame is a whole food and one cup contains seventeen grams of protein and eight grams of fiber. Easy to prepare, edamame cooks up in minutes and makes a buttery and sweet tasting snack or main vegetable dish.

Soybeans, Mature, Dried

As soybeans mature in the pod they ripen into a hard, dry seed in the field before being harvested. Soybeans are classified according to color—yellow, brown or black, with yellow soybeans being the most common.

Whole soybeans are rinsed and then soaked for eight to eighteen hours before proceeding with any recipe. Soybeans require cooking for about three hours; however, there is no need to tend them closely and a pressure cooker can reduce the cooking time to less than a half hour. One cup of dry soybeans is equal to about two and a half cups cooked.

Whole soybeans are an excellent source of protein and dietary fiber and can be used in stews, soups, and sauces. Dry soybeans are available in natural food stores and some supermarkets, sold in bulk bins or bagged. High quality non-GMO soybeans can be ordered online from Chambers Family Farms of Iowa.

Soybean Sprouts

Although not as popular or available as mung bean sprouts or alfalfa sprouts, soybean sprouts provide excellent nutrition and are a rich source of protein and vitamin C. Soybeans are easily sprouted at home in the same manner as other beans and seeds. Like other sprouts, soybean sprouts are most often used raw in salads, in soups, or stir-fried, sautéed, or baked. However, heat should be applied sparingly to avoid mushiness.

Soy Alternatives (Meat Analogs)

Meat alternatives made from soybeans contain soy protein and other ingredients, such as wheat gluten, combined to simulate the flavor and texture of various kinds of meat. These soyfoods are available refrigerated, frozen, canned or dried. Mostly, these soy alternatives can be incorporated into recipes in the same way as animal products they replace. Soy alternatives made from soybeans provide an excellent source of protein, iron, and B vitamins and are generally lower in fat; however, the nutritional values vary across the broad range of products in this category.

Soy Cheese/Yogurt/Sour Cream

There is a burgeoning dairy-free marketplace. It seems as though new products come onto the market every month—from gourmet and artisan-style cheeses, plain and flavored sour cream and cream cheeses, and regular and Greek-style yogurt made from soy and vegan cultures. After years of development, these products deliver satisfying flavor, creamy texture, and mouthfeel without the unhealthy baggage or inherent cruelty of animal products and are available in natural food stores and many supermarkets.

Soy Flour

Soy flour is made from roasted soybeans ground into a fine powder that boosts protein and brings moisture to baked goods, improving taste and texture. There are three kinds of soy flour: 1) natural or 2) full fat, which contain the natural oils found in the soybean, and 3) defatted, which has the oils removed during processing.

Soy flour of any variety will boost protein; however, defatted soy flour is an even more concentrated source of protein than full fat soy flour. Although used mainly by the food industry, soy flour can be found in natural foods stores and some supermarkets. Soy flour is gluten-free so yeast-raised breads made with soy flour are more dense in texture. Replace one third of the flour with soy flour in recipes for muffins, cakes, cookies, pancakes, and quick breads.

Soy Grits

Soy grits are similar to soy flour except that the soybeans have been toasted and cracked into coarse pieces, rather than the fine powder of soy flour. The toasting brings out the pleasant, nutty flavor. High in fiber and protein, soy grits can be eaten as a cereal with soymilk and fruit. Soy grits are also used as a substitute for some of the flour in recipes and added to rice or other grains.

Soy Lecithin

Extracted from soybean oil, lecithin is a very common ingredient in packaged foods because it is such a great emulsifier and stabilizer. It also promotes stabilization, antioxidation, crystallization, and spattering control. Lecithin is used in the home kitchen in gluten-free baking and can be found in natural food stores.

Soymilk

Soybeans, when soaked, ground fine, and strained, produce soybean milk, aka soymilk, which is one-for-one substitute for cow milk. Soymilk is an excellent source of high quality protein, B vitamins, and phytonutrients. Soymilk is now ubiquitous in natural groceries, supermarkets, and big box stores in shelf-stable aseptic containers and gable-top quart and half-gallon containers in the dairy aisle.

Soymilk, Homemade - Automatic Soymilk Makers

Drinking soymilk is the quick and easy way to get the goodness of soy into your body by replacing unhealthy cow milk in coffee, cereal, smoothies, and protein drinks. Automatic soymilk makers turn the lengthy process of making soymilk at home into a simple twenty-minute affair. Just fill the canister with water to the line, add pre-soaked beans, and press start! The cost of a half-gallon of soymilk averages from three to four dollars and if you have one of these nifty machines you will save thousands of dollars a year. It's delicious and could not be fresher.

Soy Nuts

Roasted soybeans, also known as soy nuts, are a tasty snack that is widely available plain, salted, or in a variety of other flavors—including chocolate-covered—in natural food stores and many supermarkets.

A quarter-cup of dry-roasted soy nuts contains 194 calories. Notably, a small serving of soy nuts with its relatively low calorie count contains high levels of protein, fiber, vitamins and minerals.

Soy nuts are a whole food and are similar in texture and flavor to peanuts and may be oil roasted or dry roasted sold in bulk or packaged. Roasted soybeans are also used to make soy nut butter.

Soy Nut Butter

Soy nut butter is a tasty alternative to peanut butter made from fresh roasted whole soybeans and is remarkably similar in taste and texture to peanut butter. Soy nut butter has significantly less fat and saturated fat than peanut butter, and contains seven grams of soy protein per serving. Soy nut butter is available in creamy and crunchy styles and is a popular peanut butter alternative in schools and camps across the country.

Soy Protein Concentrate

Soy protein concentrate also comes from defatted soy flakes, which contains 70 percent protein and is higher in fiber than soy protein isolate. Soy protein concentrate is often used as an ingredient in protein shakes, snack foods, energy bars, and breakfast cereals.

The isoflavone content in soy protein concentrate depends on how it is processed. If processed with water soy protein concentrate is rich in isoflavones; however, if the concentrate is processed with alcohol it will be low in isoflavones.

Soy Protein Isolate

Soy protein isolate is protein removed from defatted soy flakes and is a good source of isoflavones. It's an easy way to incorporate soy protein into the diet as many protein drinks, shakes, and energy bars popular with fitness enthusiasts contain soy protein isolate. It's bland in flavor and available as a plain powder sold in canisters in health foods stores and natural supermarkets.

Soy Sauce (Tamari - Shoyu)

Soy sauce is a generic name for dark brown liquid seasoning made from soybeans that have been fermented. Soy sauce has a salty taste, but is lower in sodium than traditional table salt. There are two main types of soy sauces: shoyu and tamari. Shoyu is a blend of soybeans and wheat. Tamari is made only from soybeans and is a byproduct of making miso. Tamari is gluten-free.

Tempeh

Tempeh (tem-pay) is a traditional soyfood from Indonesia that has a tender, chewy texture and hearty consistency made using a controlled fermenting process that binds whole soybeans into a rich cake with a nutty, smoky flavor.

Tempeh's controlled fermentation of the whole soybean makes it the whole food choice high in dietary fiber, vitamins, protein, and vitamin B12. Whole soybeans, sometimes mixed with another grain such as rice or millet, are fermented into a rich cake of soybeans with a smoky or nutty flavor. Tempeh can be marinated and grilled or added to soups and casseroles; and chili tempeh, a traditional Indonesian food, is a chunky, tender soybean cake. Tempeh is available at natural food stores and supermarkets.

Textured Soy Protein

Textured soy protein (TSP) refers to products made from defatted textured soy flour. One of the more popular brands is made by Archer Daniels Midland Company, who owns the rights to the product name "Textured Vegetable Protein" (TVP). When hydrated TVP has a chewy texture that mimics familiar traditional dishes.

Tofu

Tofu, a staple in Asia for thousands of years is made from soymilk in the same way cheese is made from milk. Fresh, hot soymilk is curdled with a coagulant, such as calcium chloride or nigari, and curdled. Soymilk is then strained to separate the curds and soy whey.

Tofu has a neutral flavor and easily absorbs the flavors of whatever it is cooked with and flavorful marinades. Sometimes called bean curd, tofu is rich in protein and B vitamins and low in sodium.

Two Styles of Tofu: Chinese and Japanese

Chinese style tofu has a firm spongy texture that makes it the best choice for grilling, stir-frying, or braising. Firm, extra firm, or super firm tofu is dense what you want for any dish where the tofu needs to stand up to handling. Firm-style tofu is higher in protein, fat, and calcium than other forms of tofu. Look for shrink-wrapped or water-packed firm tofu in the refrigerated case at natural food stores and supermarkets.

Japanese style tofu or silken tofu has a neutral flavor and smooth, silky texture and is also known as silken tofu. This type of tofu is the perfect choice to replace eggs in baking and dairy products in creamy dishes such as puddings, pie fillings, toppings, sauces, and quiche.

Silken tofu is the only choice for recipes that call for blended tofu, as Chinese style tofu is porous and spongy. The creamy texture and rich mouthfeel of silken tofu make it the dairy-free ingredient of choice to replace sour cream in dips and salad dressings. Silken tofu is available in shelf-stable aseptic packaging on the shelf and water packed in the refrigerated case at natural food stores and supermarkets.

Whole Soybeans, Canned

Whole soybeans are available dried, roasted, and also canned.

Canned soybeans are dried beans that have been cooked for convenience by the manufacturer and can be found in many markets. Both yellow and black soybeans are available canned in natural food stores and supermarkets from Westbrae and Eden Foods.

Yuba

Yuba is considered a delicacy and is made by lifting and drying the creamy, thin layer that forms on the surface of hot soymilk as it cools. Yuba is a high-protein soyfood commonly sold fresh, dried, or partly dried.

In the US, dried yuba sheets are also called dried bean curd, bean curd sheets, or bean curd skin. These thin sheets can be cut into noodles and make great gluten-free pasta. Yuba is also sold in U-shaped rolls that is called bamboo yuba or bean curd sticks and can be found in Asian groceries.

Isoflavones

What are isoflavones and why are they good for you?

Isoflavones are biologically active compounds found abundantly in soy that reportedly have many health benefits. There are three forms of isoflavone, daidzein, genistein, and glycitein.

Isoflavones are a class of phytoestrogens that have a chemical structure that resembles human estrogen but are a thousand times weaker. According to Iowa State University's Soybean Research Program, soybeans are by far the most concentrated source of isoflavones in the plant kingdom.

Soy isoflavones can bind to estrogen receptors and may be tissue-selective, which means that isoflavones mimic the effects of estrogen in some tissues and blocks the effects of estrogen in others.

Because these weaker compounds can mimic human estrogen, consuming soyfoods rich in isoflavones may result in the dramatic reduction of the levels of human estrogen in the body.

The many health benefits reportedly associated with soflavones include protection against breast and prostate cancer, heart disease, the unpleasant symptoms of menopause, and osteoporosis.

Like other antioxidants, isoflavones are produced by the soybean plant as part of their defense mechanism against insects and disease.

And like other antioxidants, isoflavones have the property to neutralize free radicals.

Free radicals are a byproduct of normal metabolism and have a role to play in fending off invaders, such as viruses and bacteria; however, free radicals can do a lot of damage. Among the three types of isoflavones, genistein generates the strongest antioxidant activity.

Soyfoods vary in their concentration of isoflavones, and the richest source of these beneficial phytoestrogens is found in traditional soyfoods, such as soymilk, tofu, tempeh, miso, edamame, and soy nuts. That is not to say that you should swear off soy burgers, soy hot dogs, or soy chorizo—they also contain isoflavones, just not as much. The very nature of the production of the clever and very tasty alternatives necessitate combining soy with other ingredients and thus the amount of soy in the product is less than in soyfoods made wholly from the soybean.

It would be difficult to consume too much isoflavone from soyfoods, unlike concentrated isoflavone pills or supplements. Most nutritionists agree that one or two servings of whole soyfoods a day will impart the benefits of soy.

Isoflavone Content of Selected Soyfoods (milligrams per 100 grams[1])

Traditional Soyfoods and Ingredients

Food	Isoflavone	mg
Bean Curd, Fermented Fermented Tofu (Condiment)	Daidzein	12.18
	Genistein	21.12
	Glycitein	2.30
	Total Isoflavones	**34.68**
Soybean, Green Edamame Raw	Daidzein	61.70
	Genistein	60.07
	Glycitein	7.07
	Total Isoflavones	**128.83**
Soybean, Green Edamame Cooked	Daidzein	7.41
	Genistein	7.06
	Glycitein	4.60
	Total Isoflavones	**17.92**
Miso	Daidzein	16.43
	Genistein	23.24
	Glycitein	3.00
	Total Isoflavones	**41.45**
Natto	Daidzein	33.22
	Genistein	37.66
	Glycitein	10.55
	Total Isoflavones	**82.29**
Okara	Daidzein	3.62
	Genistein	4.47
	Glycitein	1.30
	Total Isoflavones	**9.39**

Food	Isoflavone	Amount
Soybeans, Mature, Canned	Daidzein	26.15
	Genistein	25.15
	Glycitein	6.10
	Total Isoflavones	**52.82**
Soybeans, Mature, Cooked, without Salt	Daidzein	30.76
	Genistein	31.26
	Glycitein	3.75
	Total Isoflavones	**65.11**
Soybeans, Mature Dry Roasted Soynuts	Daidzein	62.14
	Genistein	75.78
	Glycitein	13.33
	Total Isoflavones	**148.50**
Soybeans, Mature, Raw	Daidzein	62.07
	Genistein	80.99
	Glycitein	14.99
	Total Isoflavones	**154.53**
Soy Flour (textured) TSP	Daidzein	67.69
	Genistein	89.42
	Glycitein	20.02
	Total Isoflavones	**172.55**
Soy Flour, Defatted	Daidzein	64.55
	Genistein	87.31
	Glycitein	15.08
	Total Isoflavones	**150.94**
Soy Flour, Full-Fat, Raw	Daidzein	72.92
	Genistein	98.77
	Glycitein	16.12
	Total Isoflavones	**178.10**

Soy flour, Full-Fat, Roasted	Daidzein	89.46
	Genistein	85.12
	Glycitein	16.40
	Total Isoflavones	**165.04**
Soybeans, Sprouted Cooked,	Daidzein	5.00
	Genistein	6.70
	Glycitein	0.80
	Total Isoflavones	**12.50**
Soybeans, Sprouted Raw	Daidzein	12.86
	Genistein	18.77
	Glycitein	2.88
	Total Isoflavones	**34.39**
Soy Lecithin	Daidzein	5.40
	Genistein	10.30
	Total Isoflavones	**15.70**
Soy Protein Isolate	Daidzein	30.81
	Genistein	57.28
	Glycitein	8.54
	Total Isoflavones	**91.05**
Soy Sauce from Soy and Wheat (Shoyu)	Daidzein	0.78
	Genistein	0.39
	Glycitein	0.14
	Total Isoflavones	**1.18**
Soymilk, Chocolate Enriched, Calcium Vitamins A and D	Daidzein	3.40
	Genistein	4.15
	Glycitein	0.25
	Total Isoflavones	**7.80**

Soymilk, Plain/Vanilla Fortified or Unfortified	Daidzein	4.84
	Genistein	6.07
	Glycitein	0.93
	Total Isoflavones	**10.73**
Soymilk Powder Not Reconstituted	Daidzein	40.07
	Genistein	62.18
	Glycitein	10.90
	Total Isoflavones	**109.51**
Tempeh	Daidzein	22.66
	Genistein	36.15
	Glycitein	3.82
	Total Isoflavones	**60.61**
Tempeh Burger	Daidzein	6.40
	Genistein	19.60
	Glycitein	3.00
	Total Isoflavones	**29.00**
Tempeh, Cooked	Daidzein	13.12
	Genistein	21.14
	Glycitein	1.39
	Total Isoflavones	**35.64**
Tempeh, Fried	Daidzein	32.90
	Genistein	39.90
	Total Isoflavones	**72.80**
Tofu, Azumaya Extra Firm, Cooked	Daidzein	8.00
	Genistein	12.75
	Glycitein	1.95
	Total Isoflavones	**22.70**

Tofu, Azumaya Firm, Cooked	Daidzein	12.80
	Genistein	16.15
	Glycitein	2.40
	Total Isoflavones	**31.35**
Tofu, Braised Firm	Daidzein	7.28
	Genistein	8.22
	Glycitein	1.28
	Total Isoflavones	**16.79**
Tofu, Fried	Daidzein	13.80
	Genistein	18.43
	Glycitein	2.93
	Total Isoflavones	**34.78**
Tofu, Nigari Extra Firm	Daidzein	8.23
	Genistein	12.45
	Glycitein	1.95
	Total Isoflavones	**22.63**
Tofu, Nigari Soft	Daidzein	9.49
	Genistein	11.91
	Glycitein	1.68
	Total Isoflavones	**22.61**
Tofu, Nigari Firm	Daidzein	12.31
	Genistein	16.10
	Glycitein	2.75
	Total Isoflavones	**30.41**
Tofu, Silken	Daidzein	9.15
	Genistein	8.42
	Glycitein	0.92
	Total Isoflavones	**18.04**

Tofu, Silken, Firm Mori-Nu	Daidzein	12.42
	Genistein	16.95
	Glycitein	2.40
	Total Isoflavones	**29.97**
Tofu, Silken, Soft Vitasoy	Daidzein	8.59
	Genistein	20.65
	Total Isoflavones	**29.24**
Tofu, Smoked	Daidzein	7.50
	Genistein	5.60
	Total Isoflavones	**13.10**
Tofu, Dried, Frozen (koyadofu)	Daidzein	29.59
	Genistein	51.04
	Glycitein	3.44
	Total Isoflavones	**83.20**
Yuba, Soymilk Skin Cooked	Daidzein	17.81
	Genistein	25.15
	Glycitein	2.69
	Total Isoflavones	**44.67**
Yuba, Soymilk Skin Raw	Daidzein	80.03
	Genistein	101.40
	Glycitein	15.43
	Total Isoflavones	**196.05**

Soy Products – Meat and Dairy Alternatives

Soy Bacon Bits	Daidzein	64.37
	Genistein	45.77
	Glycitein	8.33
	Total Isoflavones	**118.50**
Soy Bacon	Daidzein	2.20
	Genistein	5.66
	Glycitein	1.50
	Total Isoflavones	**9.36**
Soy Burgers, Unprepared	Daidzein	2.36
	Genistein	5.01
	Glycitein	0.55
	Total Isoflavones	**6.39**
Soy Butter, Full-Fat Worthington Foods	Daidzein	0.22
	Genistein	0.30
	Glycitein	0.05
	Total Isoflavones	**0.57**
Soy Cheese, American	Daidzein	5.75
	Genistein	8.70
	Glycitein	3.50
	Total Isoflavones	**17.95**
Soy Cheese, Cheddar	Daidzein	1.83
	Genistein	2.11
	Glycitein	2.93
	Total Isoflavones	**6.87**

Soy Cheese Fat-Free Monterey Jack	Daidzein	7.80
	Genistein	8.80
	Glycitein	2.10
	Total Isoflavones	**18.70**
Soy Cheese, Mozzarella	Daidzein	1.14
	Genistein	2.60
	Glycitein	2.28
	Total Isoflavones	**6.02**
Soy Cheese, Parmesan	Daidzein	1.50
	Genistein	0.80
	Glycitein	4.10
	Total Isoflavones	**6.40**
Soy Cheese, Swiss	Daidzein	1.80
	Genistein	4.40
	Glycitein	1.70
	Total Isoflavones	**7.90**
Soy Cheese Unspecified	Daidzein	5.79
	Genistein	11.14
	Total Isoflavones	**25.72**
Soy Chicken Nuggets Canned Prepared Worthingtn FriChik	Daidzein	4.35
	Genistein	9.35
	Glycitein	0.85
	Total Isoflavones	**12.20**
Soy Chicken Patties Morningstar Farms Chik Patties Original	Daidzein	1.80
	Genistein	2.20
	Glycitein	0.40
	Total Isoflavones	**4.40**

Soy Chips	Daidzein	26.71
	Genistein	27.45
	Total Isoflavones	**54.16**
Soy Franks Canned prepared Loma Linda BigFranks	Daidzein	1.35
	Genistein	2.05
	Glycitein	0.30
	Total Isoflavones	**3.35**
Soy Ice Cream, Tofutti Frozen Dessert	Daidzein	1.10
	Genistein	1.70
	Glycitein	0.10
	Total Isoflavones	**2.90**
Soy Hot Dog, Frozen, Unprepared	Daidzein	0.40
	Genistein	0.60
	Glycitein	0.00
	Total Isoflavones	**1.00**
Soy Noodles, Flat	Daidzein	0.90
	Genistein	3.70
	Glycitein	3.90
	Total Isoflavones	**8.50**
Soy Sausage	Daidzein	4.46
	Genistein	9.23
	Glycitein	2.30
	Total Isoflavones	**14.34**
Soy Sausage Prepared Morningstar Farms Veggie Sausage Links	Daidzein	0.75
	Genistein	2.70
	Glycitein	0.30
	Total Isoflavones	**3.75**

Soy Sausage Unprep.	Daidzein	1.18
Morningstar Farms	Genistein	2.45
Veggie Sausage Links	Glycitein	0.30
	Total Isoflavones	**3.93**
Soy Sausage Patties	Daidzein	2.00
Morningstar Farms	Genistein	2.30
Veggie Sausage Patties	Glycitein	0.30
	Total Isoflavones	**4.60**
Tofu Mayonnaise	Daidzein	5.50
	Genistein	11.30
	Total Isoflavones	**16.80**
Soy Yogurt	Daidzein	13.77
	Genistein	16.59
	Glycitein	2.80
	Total Isoflavones	**33.17**
Tofu Yogurt	Daidzein	5.70
	Genistein	9.40
	Glycitein	1.20
	Total Isoflavones	**16.30**

ACKNOWLEDGEMENTS

I would like to acknowledge authors, research scientists, experts, and medical and nutrition professionals whose interviews and publications provided critical information in the writing of this book.

James W. Anderson, MD

Dr. Anderson trained in internal medicine and endocrinology at the Mayo Clinic and is a Professor of Medicine and Clinical Nutrition at the University of Kentucky. He directs the University of Kentucky Health Management Resources Weight Management Program and is Founder and President of the Obesity Research Network, a network of leading experts who perform clinical research in the area of obesity. Dr. Anderson has published over 220 research articles and over 190 book chapters, education articles, and books.

His publications have appeared in the *New England Journal of Medicine*, *Journal of American Medical Association*, *Journal of American Dietetic Association*, *Archives of Internal Medicine*, and *American Journal of Clinical Nutrition*. Dr. Anderson's research interests include diabetes, blood lipid disorders, obesity, and nutrition. He is currently investigating novel ways to reduce blood cholesterol, the use of soy protein in diabetes, and new treatments for obesity.

Thomas M. Badger, PhD

Dr. Thomas Badger, a senior researcher at the Arkansas Children's Nutrition Center (ACNC) served as the ACNC Director for twenty years. He is a neuroendocrinoligist and nutritionist.

His research focuses on the long-term health consequences of early nutrition and diet, specifically the prevention of childhood diseases, including obesity, and adult diseases that are initiated early in life but do not typically become evident until adulthood such as breast, prostate, or colon cancer. He also has had a long-term interest in the effects of soy-based infant formulas and the potential estrogenic effects on development.

Dr. Badger is Professor of Pediatrics and Chief of the Division of Development Nutrition and his most recent research has been to assist Drs. Aline Andres, Kartik Shankar, Xiawei OU, and Terry Pivik in a series of studies on maternal programming of fetal metabolism as relates to childhood obesity and brain development. These studies are related to women's health issues, child development, and long-term health consequences.

He is currently conducting a long-term study of the growth and development of six hundred infants from roughly birth through puberty. The purpose of the Beginnings Study is to determine differences that may exist between the three modes of infant feeding—breast, cow milk infant formula, and soy formula—and what, if any, adverse effects may result from formula feeding, with specific emphasis on body composition and brain development/function like behavior, cognition, psychomotor, and language development.

Jonathan Balcombe, PhD

Jonathan Balcombe has three biology degrees, including a PhD in ethology (the study of animal behavior) from the University of Tennessee, where he studied communication in bats.

Dr. Balcombe has published over forty-five scientific papers on animal behavior and animal protection and is a best-selling author with four books. He is currently at work on a new book about the inner lives of fishes and a novel titled *After Meat*.

Formerly Senior Research Scientist with the Physicians Committee for Responsible Medicine, Dr. Balcombe is an author and currently the Department Chair for Animal Studies with the Humane Society University. A research scientist with a background in ethology, he has written many scientific papers on animal behavior and articles on humane life science education.

Neal Barnard, MD

Neal Barnard is a clinical researcher, author, and health advocate. He has been the principal investigator or co-investigator on several clinical trials investigating the effects of diet on health. Most recently, he was the principal investigator of a study on dietary interventions in diabetes, funded by the National Institutes of Health and conducted under the auspices of the George Washington University School of Medicine, in association with the University of Toronto.

Dr. Barnard is the author of dozens of publications in scientific and medical journals as well as numerous nutrition books for lay readers and is frequently called on by news programs to discuss issues related to nutrition, research issues, and other controversial areas in modern medicine. He is a frequent lecturer at scientific and lay conferences and has made presentations for the American Diabetes Association, American Public Health Association, the World Bank, the National Library of Medicine, and many other medical and scientific organizations.

Stephen Barnes, PhD

Dr. Stephen Barnes is a London Educated Biochemist with numerous awards and published research to his credit. On the Faculty of the University of Alabama at Birmingham, Dr. Barnes is a Professor of Pharmacology and Toxicology, Professor of Biochemistry and Molecular Genetics, Professor of Environmental Health Sciences, Professor of Genetics, Professor of Vision Sciences, Director, UAB Center for Nutrient-Gene Interaction, Associate Director, Purdue-UAB Botanicals Centers for Age-Related Disease, Director, Targeted Metabolomics and Proteomics Laboratory, and Senior Scientist at UAB Comprehensive Cancer Center.

T. Colin Campbell, PhD

Dr. T. Colin Campbell, a researcher, lecturer, and policy advisor in the field of diet and cancer for more than forty years is the Jacob Gould Schurman Professor Emeritus of Nutritional Biochemistry at Cornell University. He has more than seventy grant years of peer-reviewed research funding, has authored more than three hundred research papers, and is coauthor of the bestselling the book, *The China Study: Startling Implications for Diet, Weight Loss and Long-term Health*.

Dr. T. Colin Campbell is probably best known for *The China Study* and for his role as Senior Science Advisor during the formative years of the American Institute for Cancer Research (AICR) and at the World Cancer Research Fund.

AICR awarded Dr. Campbell the 1998 Lifetime Achievement Award in Cancer Research. As a result of his work in The China Project, deemed the most comprehensive study on diet and disease ever conducted, he was named to the Board of Directors of a new government agency in China responsible for developing national diet and health research and education programs.

Alison Duncan, RD, PhD

Dr. Alison Duncan is associate professor in the Department of Human Health and Nutritional Sciences at the University of Guelph in Ontario, Canada. Dr. Duncan received the Young Canadian Innovator Award in Agriculture, Food, and Human Health in 2005.

Dr. Duncan's research interests relate to the biological effects of functional foods and nutraceuticals on chronic disease through human intervention studies. Her specific interests have involved the human health effects of soy and its constituent protein and isoflavones, which have evolved to include other dietary interventions, such as pulses, that are part of the agriculture-food-human health continuum.

Other research interests include the exploration of the use of functional foods and natural health products in healthy and clinical populations. Additionally, in Dr. Duncan's capacity as Associate Director of Research for the Human Nutraceutical Research Unit, a research and educational unit within the Department of Human Health and Nutritional Sciences, she works on clinical trials to examine the human health effects of foods and natural health products.

Christopher D. Gardner, PhD

Dr. Gardner is actively involved in research and teaching and is Professor of Research Medicine at the Stanford Prevention Research Center. For the past twenty years most of Dr. Gardener's research has been focused on investigating the potential health benefits of various dietary components or food patterns.

His research focus is on dietary intervention trials designed to test the effects of food components on chronic disease risk factors including body weight, blood lipids, and inflammatory markers.

Dr. Gardener has published extensively in peer-reviewed journals, including a recent publication in the *Journal of the American Medical Association* presenting his findings from a study contrasting the weight loss effects of four diets ranging from very low to very high carbohydrate. He speaks nationally within his areas of expertise, which include: dietary approaches to improve blood lipids, weight loss diets, health benefits of vegetarian and plant-based diets, the optimal diet for cardiovascular health, dietary supplements, and complementary and alternative medicine.

Carey Gleason, PhD

Dr. Gleason is an associate professor at the University of Wisconsin School of Medicine and senior scientist, Section of Geriatrics and Gerontology, at the University of Wisconsin-Madison and

W.S. Middleton Memorial Veteran's Hospital Geriatric Research, Education, and Research Center.

Roberta S. Gray, MD

Dr. Roberta S. Gray is a pediatric nephrologist practicing in North Carolina and South Carolina. Dr. Gray completed her internship and residency in pediatrics at the University of Kansas Medical Center in Kansas City and served as chief resident there.

Dr. Gray has held academic pediatric faculty positions at Duke University, Marshall University, East Carolina University, and Carolinas Medical Center. She often prescribes vegetarian diets for her young patients.

Jill Hamilton-Reeves, PhD, RD

Dr. Hamilton-Reeves is Associate Professor in the Department of Dietetics and Nutrition at the University of Kansas. A registered and licensed dietitian certified in oncology nutrition with doctoral training in translational and clinical nutrition science, Dr. Hamilton-Reeves teaches in the graduate programs in the Department of Dietetics and Nutrition.

Her research is focused on nutrition and cancer prevention, management, and survivorship, with specific attention on genitourinary cancers.

Dr. Hamilton-Reeves has finished six marathons, a triathlon, and an in-line skating marathon. Her research interests and specializations are hormone-like dietary compounds and chronic disease prevention, soy and prostate cancer prevention, and vitamin D and medication interactions. Dr. Hamilton-Reeves lost three of her four grandparents to cancer and is particularly interested in diet-mediated cancer prevention.

Dr. Leena A. Hilakivi-Clarke

Dr. Leena A. Hilakivi-Clarke is a professor in the Department of Oncology Lombardi Comprehensive Cancer Center of Georgetown

University School of Medicine where she is also Codirector of the Division of Molecular Endocrinology, Nutrition, and Obesity.

Dr. Hilakivi-Clarke's research involves studying the role of estrogens and diet in breast cancer. Her research interests include understanding the importance of the timing of exposure to estrogens, phytoestrogens, and other dietary and nutrition components on the programming of mammary gland development and its effects on breast cancer susceptibility.

She is also active in exploring gene-nutrient interactions and their role in affecting breast cancer risk and progression.

Barbara P. Klein, PhD *(1936–2015)

Dr. Klein was Professor Emerita of Foods and Nutrition in the Department of Food Science and Human Nutrition and a member of the Division of Nutritional Sciences at the University of Illinois at Urbana-Champaign. She was also the co-director of the Illinois Soyfoods Center.

Dr. Klein's research interests were alterations in food quality that occur during storage, processing, and preparing foods for human consumption. Her recent research focused on development and evaluation of high soy protein foods such as snacks, cereals, and dairy analogs; and factors affecting phytochemical and nutrient retention in vegetables.

Dr. Klein was editor of two books, author of seven book chapters, and more than one hundred journal articles and presentations.

*It is with sadness that we note Dr. Klein's passing on February 17, 2015. We are most grateful to have had the opportunity to interview her for this book.

Stacey L. Krawczyk, MS, RD

Stacey has been a practicing registered dietitian for more than twenty years during which time she was employed as a regional nutritionist for the Illinois WIC Program for seven years.

Stacey has been the Research Dietitian for the National Soybean Research Laboratory and a culinary nutrition instructor for the campus recreation programs at the University of Illinois. Stacey's long-standing interest and area of expertise is related to school nutrition and childhood obesity prevention, and she has been a consultant to maternal and child health programs, school nutrition programs, and diabetes education and advocacy.

Mindy S. Kurzer, PhD

Dr. Mindy S. Kurzer is a nutritional scientist and Professor in the Department of Food Science and Nutrition, and she holds a joint appointment in the Division of Hematology, Oncology, and Transplantation in the Department of Medicine.

Dr. Kurzer studies dietary effects on endogenous hormones and hormone actions as a mechanism by which substances in food may prevent cancer. She performs clinical studies in healthy subjects and cancer survivors, focusing primarily on breast and prostate cancer prevention.

The diet and lifestyle interventions that Dr. Kurzer has studied include soy phytoestrogens (isoflavones), soy protein, flaxseed, omega-3 fatty acids, and exercise.

Dr. Kurzer received the International Life Sciences Institute Future Leader Award in 1992 and was a Committee on Institutional Cooperation Academic Leadership Fellow from 2006 to 2007. She currently serves on the NCCAM Clinical Sciences study section and regularly reviews for both the Department of Defense Breast and Prostate Cancer grant programs.

Mark Messina, PhD, MS

Dr. Mark Messina is an internationally recognized expert on the health effects of soy. Dr. Messina has presented to both consumer and professional audiences in forty-four countries on topics including soy and cancer risk, heart health, menopause, and men's health.

Mark is the coauthor of *The Simple Soybean and Your Health*. His research has appeared in numerous professional journals including the *American Journal of Clinical Nutrition*, *Journal of Nutrition*, *Lancet*, and the *Journal of the National Cancer Institute*.

Dr. Messina is a former program director in the Diet and Cancer Branch, National Cancer Institute, National Institutes of Health, where he initiated a multimillion-dollar research program investigating the role of soy in cancer prevention.

Mark co-owns Nutrition Matters, Inc., a nutrition consulting company, is an adjunct associate professor in the Department of Nutrition, School of Public Health, Loma Linda University, and is the executive director of the Soy Nutrition Institute.

Dixie Mills, MD

Dixie Mills is a renowned breast care specialist and Harvard-trained surgeon, specializing in breast care since 1989. Dr. Mills practices a holistic approach to medicine and wellness and has lectured on a wide variety of topics, including "Complementary Breast Cancer Treatments" and "Redefining the Doctor/Patient Relationship."

Dr. Mills has served as Chief Surgical Resident at Deaconess Hospital, Harvard Medical School, is certified by the American Board of Surgery, and is a Fellow of the American College of Surgeons. She is also a member of the Association of Women Surgeons and the National Breast Cancer Coalition.

Dr. Mills cofounded the Breast Cancer High-Risk Clinic at the Dana-Farber Cancer Institute in Boston and served as the Clinical Research Director for Dr. Susan Love's Research Foundation in California and was Medical Director from 2008 until May of 2010.

John Robbins

John Robbins is the author of many best selling books, including *The Food Revolution*, *The New Good Life*, and the classic *Diet for a New America*. John is considered to be one of the world's leading experts on the dietary link between health and the environment.

Robbins is an eloquent spokesperson for a healthy and sustainable future, and has received standing ovations at thousands of conferences and speaking engagements worldwide, including at the United Nations.

He is the founder of EarthSave International, a nonprofit organization that supports healthy food choices, preservation of the environment, and a more compassionate world. Many leading authorities in health and ecology have called John Robbins's work among the most important of the century.

John is the recipient of the 1994 Rachel Carson Award, the Albert Schweitzer Humanitarian Award, the Peace Abbey's Courage of Conscience Award, and Green America's Lifetime Achievement Award. His life and work have been featured in the PBS special *Diet for a New America.*

Lawrence Ross, MD

Lawrence Ross is the Clarence C. Saelhof Professor Emeritus for the Department of Urology at University of Illinois at Chicago, where he has worked for twenty-five years. Previously, Dr. Ross served as the Clarence C. Saelhof Professor and Head of the Department of Urology.

Dr. Ross is an internationally renowned urologist focusing on male reproduction, infertility surgery, and research. He has studied sperm cryopreservation and varicocele surgery, including the effects of varicocele on adolescents in order to determine the optimum time of repair for young men. Dr. Ross is presently the American Urological Association liaison for the Centers for Disease Control and Prevention.

Haley Curtis Stevens, PhD

Haley Curtis Stevens was the Scientific Affairs Specialist for the International Formula Council, a North American trade association of infant formula manufacturers and adult nutritionals.

In that capacity, Dr. Stevens provided scientific expertise on various issues involving infant formulas (e.g., melamine, BPA, perchlorate) as well as worked on the Codex Commission on Nutrition and Foods for Special Dietary Uses (CCNFSDU) methods of analysis for infant formula.

Prior to joining IFC, Dr. Stevens completed an undergraduate degree at Vanderbilt University in Nashville and a doctorate in Biological and Biomedical Sciences at Emory University in Atlanta.

William Shurtleff

William Shurtleff is a best-selling author and founder the SoyInfo Center, which developed SoyaScan. Developed in 1985, SoyaScan is a computerized bibliographic database and the world's most comprehensive source of information on all aspects of soy.

Bill Shurtleff was born in Oakland, California in 1941 and he grew up on a two-and-a-half-acre mini-farm on the edge of Lafayette, California. Bill graduated from Stanford University in 1963 and joined the Peace Corps where he spent the next two years teaching high school physics in Nigeria.

Shurtleff has lived and traveled extensively in Japan and is considered a foremost expert on soy. He has authored three books with his wife, Akiko Aoyagi, *The Book of Tofu*, *The Book of Miso*, and *The Book of Tempeh*.

Jonathan Chambers

Jonathan is a fifth-generation farmer. His father, Norman; grandfather, Earl; great-grandfather, Harry; and great-great-grandfather, Samuel, all worked the land before him.

Jonathan graduated from Iowa State University with a degree in agribusiness and a vocal minor. Jonathan is very active in community theater and has four children with his wife, Paula.

Jonathan has developed a strain of non-GMO soybeans that make pearly white soymilk without a hint of beanyness. Laura® Soybeans are bred the old-fashioned way, through natural crossbreeding and are the basis for his line of Tosteds® dry roasted soybeans.

ABOUT THE AUTHOR

Marie Oser is a best-selling author, healthy lifestyle expert, and environmental advocate with a focus on nutrition and its role in disease prevention. Vegetarian since 1971 and vegan since 1990, Marie left a career in television advertising to pursue her interest in food, health, nutrition, and the environment.

She began writing for major metros and national magazines in 1990 and was a nationally syndicated newspaper columnist, features writer, and regular contributor to publications such as *Vegetarian Gourmet, Vegetarian Times, Great Life, Natural Health, Today's Dietician,* and *Spa* magazine.

Marie is Host and Executive Producer of VegTV (www.vegtv.com). VegTV has been producing content since early 2001 and currently hosts 800 original videos on 9 channels featuring popular cooking segments and interviews with celebrities, prominent authors, and medical and nutrition experts. She's also Managing Editor at Ecomii (www.ecomii.com), a leading environmental site. At the Food & Health Alternative, Oser and her team of experts report on creative natural solutions for issues affecting health and well-being and cutting edge strategies for optimal health.

Marie lives in Southern California. *The Skinny on Soy* is her fifth book.

NOTES

Introduction

1. Wong WW, Smith EO, Stuff JE, Hachey DL, Heird WC, Pownell HJ. "Cholesterol-lowering effect of soy protein in normocholesterolemic and hypercholesterolemic men." *Am J Clin Nutr*. 1998 Dec;68 (6 Suppl):1385S–1389S.
2. Kurowska EM, Jordan J, Spence JD, et al. "Effects of substituting dietary soybean protein and oil for milk protein and fat in subjects with hypercholesterolemia." *Clin Invest Med*. 1997;20:162–170.
3. Baum JA, Teng H, Erdman JW Jr, et al. "Long-term intake of soy protein improves blood lipid profiles and increases mononuclear cell low-density-lipoprotein receptor messenger RNA in hypercholesterolemic, postmenopausal women." *Am J Clin Nutr*. 1998;68:545–551.
4. Lenka Varinska 1, Peter Gal, Gabriela Mojzisova, Ladislav. Mirossay, Jan Mojzis. "Soy and Breast Cancer: Focus on Angiogenesis Int. J. Mol." *Sci*. 2015, 16, 11728–11749.

Chapter One: Soyfoods on the Menu

1. "Soyfoods Come of Age," Richard Leviton, *Vegetarian Times*, November, 1982 http://bit.ly/1PdA4Nf.
2. "America's Milk Business in a Crisis Ian Berry," Kelsey Gee, Dec. 11, 2012, wsj.com/articles/SB10001424127887323316804578165503947704328.
3. "America's Move to Soy Hobbles Dairy," Jacob Bunge, David Kesmodel, *Wall Street Journal,* July 18, 2014, wsj.com/articles/americas-move-to-soy-hobbles-dairy-1405729869._
4. "Milk sales continue to slide as diets, society shift away from dairy." Eric Atkins, *The Globe and Mail*, Aug. 26, 2015.

Chapter Two: Fact, Fiction, and Fallacy

1. Fallon, Enig, "Tragedy and Hype." *Nexus Magazine*, volume 7, no. 3, April-May 2000.

2. *Nourishing Traditions* by Sally Fallon, Mary Enig, PhD New Trends Publishing.
3. *The China Study*, © Dr. T. Colin Campbell, Thomas M. Campbell, MD, 2005.
4. Brody, Jane E. "Huge Study Of Diet Indicts Fat And Meat", *The New York Times*, May 8, 1990.
5. *The China Study*, Dr. T. Colin Campbell, Thomas M. Campbell, MD, 2005.
6. Brody, Jane E. "Huge Study Of Diet Indicts Fat And Meat", *The New York Times*, May 8, 1990.
7. "What is campylobacteriosis?" Centers for Disease Control (CDC), Food Safety - Food Bourne Germs and Illnesses, cdc.gov/nczved/divisions/dfbmd/diseases/campylobacter.
8. Longenberger AH, Palumbo AJ, Chu AK, Moll ME, Weltman A, Ostroff SM. Campylobacter jejuni infections associated with unpasteurized milk—multiple states, 2012. Clin Infect Dis 2013;57:263–6.
9. "Notes from the Field: Recurrent Outbreak of Campylobacter jejuni Infections Associated with a Raw Milk Dairy — Pennsylvania, April–May 2013." Centers for Disease Control (CDC), Morbidity and Mortality Weekly Report (MMWR) cdc.gov/mmwr/preview/mmwrhtml/mm6234a4.htm.
10. "Health Officials Say Raw Milk Probably Caused Campylobacter Outbreak at Wisconsin High School." *Food Safety News*, News Desk, October 24, 2014 foodsafetynews.com/2014/10/health-officials-say-raw-milk-probably-caused-campylobacter-out break-at-wisconsin-school/#.VoMWqcArLEU.
11. "2015 Outbreak of Campylobacter jejuni linked to Claravale Farm brand raw milk, California" Foodbourne Illness Outbreak Database, Marler Clark. http://outbreakdatabase.com/details/2015-outbreak-of-campylobacter-jejuni-linked-to-claravale-farm-brand-raw-milk-california/.
12. Hu FB ; Manson JE ; "Willett WC Types of dietary fat and risk of coronary heart disease: a critical review." *J Am Coll Nutr*. 2001; 20(1):5-19.
13. "Meat, Organs, Bones and Skin." Christopher Masterjohn, July 2, 2013 Nutrition for Mental Health, The Weston A. Price Foundation, westonaprice.org/health-topics/meat-organs-bones-and-skin/.
14. "Beyond Cholesterol." Christopher Masterjohn, January 20, 2014 Modern Diseases. The Weston A. Price Foundation, westonaprice.org/modern-diseases/beyond-cholesterol/.
15. Author Archives: Christopher Masterjohn. The Weston A. Price Foundation westonaprice.org/author/cmasterjo/page/2/.
16. *The China Study: The Most Comprehensive Study of Nutrition Ever Conducted and the Startling Implications for Diet, Weight Loss and Long-term Health.* Thomas Campbell, T. Colin Campbell, 2006. BenBella Books.
17. Junshi Chen; T. Colin Campbell; Junyao Li; R. Peto (1990). *Diet, lifestyle, and mortality in China: a study of the characteristics of 65 Chinese counties.* Oxford, UK; Cornell, USA; Beijing, PRC: Oxford University Press; Cornell University Press; People's Medical Publishing House.
18. Junshi Chen; Banoo Parpia; T. Colin Campbell (1998). "Diet, lifestyle, and the etiology of coronary artery disease: the Cornell China Study". *The American Journal of Cardiology* (Elsevier Science) 82 (10 Supplement 2).
19. Rodricks, J. *Mycotoxins in Human and Animal Health*. Pathotox Publishers, Inc. Park Forest South, IL. 1977.

20. "Natural co-occurrence of fumonisins, deoxynivalenol, zearalenone and aflatoxins in field trial corn in Argentina." *Fod Addit Contam* 1999 Dec;16(12):565–9.
21. Walker, SJ. "Warning letter to Joseph Mercola," D.O., Feb 16, 2005, casewatch.org/fdawarning/prod/2005/mercola.shtml.
22. MacIntire, SJ. "Warning letter to Joseph Mercola," D.O., September 21, 2006.
23. Silverman, S "Warning letter to Joseph Mercola," D.O., March 22, 2011.

Chapter Three: Rumor: Soy Is an Incomplete Protein

1. FAO/WHO. *Protein Quality Evaluation Report of Joint FAO/WHO Expert Consultation*. Rome: Food and Agriculture Organization of the United Nations, 1991.
2. Havala, S. and Dwyer, J. (1988). "Position of the American Dietetic Association: vegetarian diets - technical support paper," *J. Am. Diet*. Assn., 88, 352-355.
3. "Meta-Analysis of the Effects of Soy Protein Intake on Serum Lipids."Anderson, James W., Johnstone, Bryan M, Cook-Newell, Margaret E., *N Engl J Med* 1995; 333:276-282 August 3, 1995.
4. Knight EL, Stampfer MJ, Hankinson SE, Spiegelman D, Curhan GC. "The impact of protein intake on renal function decline in women with normal renal function or mild renal insufficiency." *Ann Int Med*. 2003;138:460–467.
5. Breslau NA, Brinkley L, Hill KD, Pak CYC. "Relationship of animal protein-rich diet to kidney stone formation and calcium metabolism." *J Clin Endocrinol*. 1988;66:140–146. k-state.edu/paccats/Contents/Nutrition/PDF/Needs.pdf.

Chapter Four: Rumor: Heart Health Benefits of Soy in Question

1. Brown L, Rosner B, Willett WW, Sacks FM. "Cholesterol-lowering effects of dietary fiber: a meta-analysis." *Am J Clin Nutr*. 1999 Jan;69(1):30-42.
2. Fuchs CS, Giovannucci EL, Colditz GA, et al. "Dietary fiber and the risk of colorectal cancer and adenoma in women." *N Engl J Med*. 1999;340:169-176.
3. Soler M, Bosetti C, Franceschi S, et al. "Fiber intake and the risk of oral, pharyngeal and esophageal cancer." *Int J Cancer*. 2001;91:283-287.
4. Marlett JA, Hosig KB, Vollendorf NW, Shinnick FL, Haack VS, Story JA. "Mechanism of serum cholesterol reduction by oat bran." *Hepatology*. 1994;20:1450-1457.
5. Marlett JA, Hosig KB, Vollendorf NW, Shinnick FL, Haack VS, Story JA. "Mechanism of serum cholesterol reduction by oat bran." *Hepatology*. 1994; 20:1450-1457.
6. James W. Anderson, M.D., Bryan M. Johnstone, PhD, and Margaret E. Cook-Newell, M.S., R.D. N Engl J Med 1995; 333:276-282.

Chapter Five: Rumor: Soy Contains Dangerous Substances

1. "Tragedy and Hype," S. Fallon, M. Enig, *Nexus Magazine* 7(3) April-May 2000.
2. Prof Kai-Uwe Eckardt, MD, Prof Josef Coresh, MD, Prof Olivier Devuyst, MD, Prof Richard J Johnson, MD, Anna Köttgen, MD, Prof Andrew S Levey, MD, Prof Adeera Levin, MD. "Evolving importance of kidney disease: from subspecialty to global health burden." *Lancet*. Volume 382, No. 9887, p158–169, 13 July 2013.

3. Navaneethan SD, Jolly SE, Schold JD, Arrigain S, Saupe W, Sharp J, Lyons J, Simon JF, Schreiber MJ Jr, Jain A, Nally JV Jr. "Development and validation of an electronic health record-based chronic kidney disease registry." *Clin J Am Soc Nephrol.* 2011 Jan;6(1):40-9.
4. Chronic kidney disease/vegetarian diet, Eric Castle, MD Mayo Clinic Online. Accessed Oct. 14, 2015, mayoclinic.com/health/renal-diet/AN01465.
5. Sotelo-López A, Hernández-Infante M, Artegaga-Cruz ME. "Trypsin inhibitors and hemagglutinins in certain edible leguminosae." *Arch Invest Med (Mex).*1978;9(1):1–14.
6. Lestienne I, Icard-Vernière C, Mouquet C, Picq C, Trèche S. "Effects of soaking whole cereal and legume seeds on iron, zinc and phytate contents." *Food Chemistry*, 89(3) 421–425 February 2005.
7. Beecher GR "Overview of dietary flavonoids: nomenclature, occurrence and intake". *J. Nutr.* 133 (10): 3248S–3254S. October 2003.
8. Chung, K.T.; Wei, C.I.; Johnson, M.G. (1998). "Are tannins a double-edged sword in biology and health?" *Trends in Food Science & Technology* 9 (4): 168–175.
9. Kuiper GG, Carlsson B, Grandien K, Enmark E, Haggblad J, Nilsson S, Gustafsson JA. "Comparison of the ligand binding specificity and transcript tissue distribution of estrogen receptors alpha and beta." *Endocrinology* 1997;138:863-870.
10. Zubik L, Meydani M. "Bioavailability of soy bean isoflavones from aglycone and glucoside forms in American women." *Am J Clin Nutr* 2003;77:1459-1465.
11. Setchell KD, Brown NM, Zimmer-Nechemias L, Brashear WT, Wolfe BE, Kirschner AS, Heubi JE. "Evidence for lack of absorption of soy isoflavone glycosides in humans, supporting the crucial role of intestinal metabolism for bioavailability." *Am J Clin Nutr* 2002;76:447-453.
12. Marounek M, Duskova D, Skrivanova V. "Hydrolysis of phytic acid and its availability in rabbits." *Br J Nutr* 2003;89:287–294.
13. Derman DP, Ballot D, Bothwell TH, MacFarlane BJ, Baynes RD, MacPhail AP, Gillooly M, Bothwell JE, Bezwoda WR, Mayet F. "Factors influencing the absorption of iron from soya-bean protein products." *Br J Nutr* 1987;57:345–353.
14. Wada K, Tsuji M, Tamura T, Konishi K, Kawachi T, Hori A, Tanabashi S, Matsushita S, Tokimitsu N, Nagata C. "Soy isoflavone intake and stomach cancer risk in Japan: From the Takayama study." *Int J Cancer.* 2015 Aug 15;137(4):885–92.
15. "Tragedy and Hype," S. Fallon, M. Enig, *Nexus Magazine* 7(3) April-May 2000.
16. "Soy Alert - Tragedy and Hype." Sally Fallon and Mary G. Enig, PhD, *Nexus Magazine.* 2000 April-May; 7(3).
17. "Technology of Production of Edible Flours and Protein Products from Soybeans." Agriculture/Consumer Protection, fao.org/docrep/t0532e/t0532e02.htm.
18. Heaney RP, Weaver CM, Fitzsimmons ML. "Soybean phytate content: effect on calcium absorption." *Am J Clin Nutr* 1991; 53:745-7.
19. Weaver CM, Plawecki KL. "Dietary calcium: adequacy of a vegetarian diet." *Am J Clin Nutr* 1994; 59:1238S-1241S.
20. Arjmandi BH, Smith BJ. "Soy isoflavones' osteoprotective role in postmenopausal women: mechanism of action." *J Nutr Biochem* 2002; 13:130-137.
21. Breslau NA, Brinkley L, Hill KD, Pak CY. "Relationship of animal protein-rich diet to kidney stone formation and calcium metabolism." *J Clin Endocrinol Metab* 1988; 66:140-6.
22. Messina VK, Burke KI. "Position of the American Dietetic Association: vegetarian diets." *J Am Diet Assoc* 1997; 97:1317–21.

23. Ma J, Stampfer MJ. "Body iron stores and coronary heart disease." *Clin Chem* 2002; 48:601–3.

Chapter Six: Rumor: Eating Soy Causes Thyroid Problems

1. McCarrison R. "The goitrogenic action of soya-bean and ground-nut." *Ind J Med Res* 1933; XXI:179–181.
2. Sharpless GR, Pearsons J, Prato GS. "Production of goiter in rats with raw and treated soy bean flour." *J Nutr*, 1939;17:545–55.
3. Halverson AW, Zepplin M, Hart EB. "Relation of iodine to the goitrogenic properties of soybeans." *J Nutr*, 1949;38:115–28.
4. Wilgus HS, Gassner FX, Patton AR, Gustavson RG. "The goitrogenicity of soybeans." *J Nutr*, 1941;22:43-52.
5. Bruce B, Messina M, Spiller GA. "Isoflavone supplements do not affect thyroid function in iodine-replete postmenopausal women." *J Med Food.* 2003 Winter;6(4):309–16.
6. Ladenson PW, Singer PA, Ain KB, et al. "American Thyroid Association guidelines for detection of thyroid dysfunction." *Arch Intern Med* 2000; 160:1573–5.

Chapter Seven: Rumor: Soy Formula Is Dangerous for Infants

1. Van Wyk JJ, Arnold MB, Wynn J, Pepper F. "The effects of a soybean product on thyroid function in humans." *Pediatrics*, 1959;24:752–60.
2. Strom B L., Schinnar R, Ziegler E E., Barnhart K T., Sammel M D, Macones G A, Stallings V A, Drulis J M, Nelson S E, Hanson S A. "Exposure to Soy-Based Formula in Infancy and Endocrinological and Reproductive Outcomes in Young Adulthood."-*JAMA.* 2001;286(7):807–814.
3. Ruhrah J. "The soybean in infant feeding: preliminary report." *Arch Pediatr* 1909; 26:496–501.
4. Karjalainen J, Martin JM, Knip M, Ilonen J, Robinson BH, Savilahti E, Akerblom HK, Dosch HM. "A bovine albumin peptide as a possible trigger of insulin-dependent diabetes mellitus." *N Engl J Med.* 1992 Jul 30;327(5):302-7.
5. Gilchrist JM, Moore MB, Andres A, Estroff JA, Badger TM. "Ultrasonographic patterns of reproductive organs in infants fed soy formula: comparisons to infants fed breast milk and milk formula." *J Pediatr* 2010; 156:215–20.
6. Andres A, Cleves MA, Bellando JB, Pivik RT, Casey PH, Badger TM. "Developmental Status of 1-Year-Old Infants Fed Breast Milk, Cow's Milk Formula, or Soy Formula." *Pediatrics* 2012; 112:991–95.
7. Andres A, Casey PH, Cleves MA, Badger TM. "Body fat and bone mineral content of infants fed breast-milk, cow's-milk formula or soy formula during the first year of life." *J Pediatr.* 2013;163:143–169.
8. "Ultrasonographic Patterns of Reproductive Organs in Infants Fed Soy Formula: Comparisons to Infants Fed Breast Milk and Milk Formula." Gilchrist, JM, Moore, MB, Andres, A, Estroff, JA, Badger, TM. *JPediatrics* February 2010 156;(2) 215–220.
9. Bhatia J, Greer F. "Use of soy protein-based formulas in infant feeding. American Academy of Pediatrics Committee on Nutrition." *Pediatrics.* 2008 May;121(5):1062–8.

Chapter Eight: Rumor: Soy Has a Feminizing Effect on Males

1. Hamilton-Reeves JM, Vazquez G, Duval SJ, Phipps WR, Kurzer MS, Messina MJ. "Clinical studies show no effects of soy protein or isoflavones on reproductive hormones in men: results of a meta-analysis." *Fertil Steril.* 2010 Aug;94(3):997–1007.
2. Hamilton-Reeves JM, Rebello SA, Thomas W, Slaton JW, Kurzer MS. "Isoflavone-Rich Soy Protein Isolate Suppresses Androgen Receptor Expression without Altering Estrogen Receptor-{beta} Expression or Serum Hormonal Profiles in Men at High Risk of Prostate Cancer." *J Nutr* 2007;137:1769-75.
3. Chavarro J.E., Toth T.L., Sadio S.M., Hauser R. "Soyfood and soy isoflavone intake in relation to semen quality parameters among men from an infertility clinic." *Hum Reprod* 2008; 23(11): 2584–2590.
4. Mínguez-Alarcón L, Afeiche MC, Chiu YH, Vanegas JC, Williams PL, Tanrikut C, Toth TL, Hauser R, Chavarro JE. "Male soyfood intake was not associated with in vitro fertilization outcomes among couples attending a fertility center." *Andrology.* 2015 Jul; 3(4):702–8.
5. "Is This the Most Dangerous Food for Men?" Jim Thornton. Men's Health Magazine, May 19, 2009, menshealth.com/nutrition/soys-negative-effects.
6. Messina, M., PhD. "Soybean isoflavone exposure does not have feminizing effects on men: a critical examination of the clinical evidence Fertil Steril." May 1, 2010; (93); 7:2095–2104.
7. Yan L, Spitznagel EL. "Soy consumption and prostate cancer risk in men: a revisit of a meta-analysis." *Am J Clin Nutr.* 2009;89:1155–1163.
8. C. Munro, M. Harwood., J. J. Hlywka, et al., "Soy Isoflavones: A Safety Review," *Nutr Rev* 61 (2003): 1–33.

Chapter Nine: Rumor: Eating Tofu Causes Alzheimer's Disease

1. White LR, Petrovitch H, Ross GW, et al. "Brain aging and midlife tofu consumption." *J Am Coll Nutr* 2000; 19:242–55.
2. US Food and Drug Administration (FDA) Department of Health and Human Services Subchapter B—Food For Human Consumption CFR - Code of Federal Regulations Title 21, accessdata.fda.gov/scripts/cdrh/cfdocs/cfcfr/cfrsearch.cfm?fr=101.82
3. "Is It Safe to Eat Soy?" Messina V, MPH, RD Messina M, PhD, veganhealth.org/articles/soymessina#fn12.
4. Brzezinski A, Debi A. "Phytoestrogens: the 'natural' selective estrogen receptor modulators?" *Eur J Obstet Gynecol Reprod Biol* 1999; 85:47–51.
5. In statistics, a confounding variable or confounder is an extraneous variable that can include subject variables such as, age, gender, health status, or background.
6. Giem, P., W. L. Beeson, and G.E. Fraser. 1993. "The Incidence of Dementia and Intake of Animal Products: Preliminary Findings from the Adventist Health Study." *Neuroepidemiology* 12: 28–36.
7. Fraser G E, Singh P N, Bennett H. "Variables Associated with Cognitive Function in Elderly California Seventh day Adventists." *Am. J. Epidemiol.* 1996;143: 1181–1190.
8. File, S.E., Jarrett, N., Fluck, E., Duffy, R., Casey, K. & Wiseman, H. (2001) "Eating soya improves human memory." *Psychopharmacology* 157: 430–436.

9. Duffy, R., Wiseman, H. & File, S.E. (2003) "Improved cognitive function in postmenopausal women after 12 weeks of consumption of a soya extract containing isoflavones." *Pharmacol Biochem Behav* (in press).
10. Kritz-Silverstein, D., Von Muhlen, D. & Barrett-Connor, E. (2001) "The soy and post-menopausal health in aging (SOPHIA) study: overview and baseline cognitive function." *Abstracts of 4th Int Symposium on the Role of Soy*, San Diego.
11. Giem P, Beeson WL, Fraser GE. "The incidence of dementia and intake of animal products: preliminary findings from the Adventist Health Study." *Neuroepidemiology* 1993; 12:28–36.
12. Fraser GE. "Associations between diet and cancer, ischemic heart disease, and all-cause mortality in non-Hispanic white California Seventh-day Adventists." *Am J Clin Nutr.* 1999 Sep;70(3 Suppl):532S–538S.

Chapter Ten: Rumor: Eating Soy Causes Cancer

1. Srinivas J. Rayaprolu, Navam S. Hettiarachchy, Pengyin Chen, Arvind Kannan and Andronikos Mauromostakos. "Peptides derived from high oleic acid soybean meals inhibit colon, liver and lung cancer cell growth," Food Research International, 2013 Jan; 50(1):282–288.
2. Fotsis, Theodore; Pepper, M; Aldercreutz, H;Hase, T.; Montesano, R. ;Schweigerer, L. 1995 "Genistein, a dietary ingested isoflavanoid, inhibits cell proliferation and in vitro angiogenesis and angiogenic diseases." *Journal of Nutrition.* 125(3S):800S March. Supplement.
3. Cao C, Li SR, Dai X, Chen YQ, Feng Z, Qin X, Zhao Y, Wu J. "The effects of genistein on tyrosine protein kinase-mitogen activated protein kinase signal transduction pathway in hypertrophic scar fibroblasts." *Zhonghua Shao Shang Za Zhi.* 2008 Apr;24(2):118–21.
4. E. Matos and A. Brandani, "Review on meat consumption and cancer in South America," *Mutation Research*, vol. 506-507, pp. 243–249, 2002.
5. C. Y. Tsai, Y. H. Chen, Y. W. Chien, W. H. Huang, and S. H. Lin, "Effect of soy saponin on the growth of human colon cancer cells," *World Journal of Gastroenterology*, vol. 16, no. 27, pp. 3371–3376, 2010.
6. Hsu, Anna, Bray, Tammy M, Helferich, William G, Doerge, Daniel R, Ho, Emily. "Differential effects of whole soy extract and soy isoflavones on apoptosis in prostate cancer cells." *Exp. Biol. Med.* 2010 235: 90–97
7. Xu L, Ding Y, Catalona WJ, Yang XJ, Anderson WF, Jovanovic B, Wellman K, Killmer J, Huang X, Scheidt KA, Montgomery RB, Bergan RC. "MEK4 function, genistein treatment, and invasion of human prostate cancer cells." *J Natl Cancer Inst.* 2009 Aug 19;101(16):1141–55.
8. Yishuo Wu, Limin Zhang, Rong Na, Jianfeng Xu, Zuquan Xiong, Ning Zhang, Wanjun Dai, Haowen Jiang, and Qiang Ding. "Plasma genistein and risk of prostate cancer in Chinese population." *Int Urol Nephrol.* 2015; 47(6): 965–970.
9. Gregory B. Lesinski,, Patrick K. Reville, Thomas A. Mace, Gregory S. Young, Jennifer Ahn-Jarvis, Jennifer Thomas-Ahner, Yael Vodovotz, Zeenath Ameen, Elizabeth M. Grainger, Kenneth Riedl, Steven J. Schwartz, and Steven K Clinton. "Consumption of soy isoflavone enriched bread in men with prostate cancer is associated with reduced pro-inflammatory cytokines and immunosuppressive cells," http://cancerprevention research.aacrjournals.org/content/early/2015/08/13/1940-6207.CAPR-14-0464.abstract.

10. *The New Puberty: How to Navigate Early Development in Today's Girls*, Rodale, Louise Greenspan, Julianna Deardorff, 2014.
11. Jiang, X., Patterson, N.M., Ling, Y., Xie, J., Helferich, W.G., and Shapiro, D.J., (2008). "Low concentrations of the soy phytoestrogen genistein induce proteinase inhibitor 9 and block killing of breast cancer cells by immune cells." *Endocrinology*. 149(11): p. 5366–73.
12. "Does high dietary soy intake affect a woman's risk of primary or recurrent breast cancer?" *J Fam Pract*. 2015 October;64(10):660–662.
13. Shu XO, Zheng Y, Cai H, et al. "Soyfood intake and breast cancer survival." *JAMA*. 2009;302:2437-2443. Ballard-Barbash R, Neuhouser ML. "Challenges in design and interpretation of observational research on health behaviors and cancer survival." *JAMA*. 2009;302:2483-2484.
14. Leena Hilakivi-Clarke, Juan E. Andrade and William Helferich. "Is Soy Consumption Good or Bad for the Breast?" *J Nutr*. 2010 Dec; 140(12): 2326S–2334S.
15. Youngjoo Kwon. "Effect of soy isoflavones on the growth of human breast tumors: findings from preclinical studies." *Food Sci Nutr*. 2014 Nov; 2(6): 613–622.
16. Ballard-Barbash R, Neuhouser ML. "Challenges in design and interpretation of observational research on health behaviors and cancer survival," *JAMA*. 2009;302:2483–2484.
17. R. M. Harris, D. M. Wood, L. Bottomley, S. Blagg, K. Owen, P. J. Hughes, R. H. Waring, and C. J. Kirk, "Phytoestrogens Are Potent Inhibitors of Estrogen Sulfation: Implications for Breast Cancer Risk and Treatment," *J Clin Endocrinol Metab*. 2004 Apr;89(4):1779-87.
18. Caan BJ, Natarajan L, Parker BA, et al. "Soyfood Consumption and Breast Cancer Prognosis." *Cancer Epidemiol Biomarkers Prev*. 2011;20(5):854–858.
19. Shu XO, Zheng Y, Cai H, Gu K, Chen Z, Zheng W, Lu W. "Soyfood intake and breast cancer survival." *JAMA*. 2009 Dec 9;302(22):2437–43.
20. Messina, M, Wu, A. "Perspectives on the soy-breast cancer relation." *Am J Clin Nutr* May 2009 vol. 89 no. 5 1673S-1679S.
21. Shon YH, Park SD, Nam KS. "Effective chemopreventive activity of genistein against human breast cancer cells." *J Biochem Mol Biol*. 2006 Jul 31;39(4):448–51.
22. Shon YH, Park SD, Nam KS. "Effective chemopreventive activity of genistein against human breast cancer cells." *J Biochem Mol Biol*. 2006 Jul 31;39(4):448–51.
23. Shu XO, Zheng Y, Cai H, et al. "Soyfood intake and breast cancer survival." *JAMA*. 2009;302:2437–2443.
24. Ballard-Barbash R, Neuhouser ML. "Challenges in design and interpretation of observational research on health behaviors and cancer survival." *JAMA*. 2009;302:2483-2484.
25. Vogel VG, Costantino JP, Wickerham DL, et al. Effects of tamoxifen vs raloxifene on the risk of developing invasive breast cancer and other disease outcomes: the NSABP Study of Tamoxifen and Raloxifene (STAR) P–2 trial. JAMA 2006; 295(23):2727–2741.
26. Fisher B, Costantino JP, Wickerham DL, et al. Tamoxifen for prevention of breast cancer: report of the National Surgical Adjuvant Breast and Bowel Project P–1 Study. Journal of the National Cancer Institute 1998; 90(18): 1371–138.
27. Tamoxifen for early breast cancer: an overview of the randomised trials. Early Breast Cancer Trialists' Collaborative Group. Lancet 1998; 351(9114):1451–1467.

28. Gorin MB, Day R, Costantino JP, et al. Long-term tamoxifen citrate use and potential ocular toxicity. American Journal of Ophthalmology 1998; 125(4):493–501.
29. Cleveland Clinic, Tamoxifen and Breast Cancer. Accessed October 15, 2015. https://my.clevelandclinic.org/health/drugs_devices_supplements/hic_Tamoxifen_and_Breast_Cancer.
30. Sally Fallon and Mary G. Enig, PhD. "Tragedy and Hype: The Third International Soy Symposium," July, August/September TLfDP, 2000.
31. W. L. Spangler, M. R. Gumbmann, I. E. Liener, J. J. Rackis. "The USDA trypsin inhibitor study. III. Sequential development of pancreatic pathology in rats." *Qualifications in Plant Foods in Human Nutrition* 1985 vol. 35:232.
32. Reinwald S, Akabas SR, Weaver CM. "Whole versus the piecemeal approach to evaluating soy." *J Nutr.* 2010 Dec;140(12):2335S–2343S.

Chapter Eleven: Rumor: Soy Causes Bone Deficiencies

1. Weston A. Price Foundation (WAPF) brochure; "Dietary Dangers," rejoiceinlife.com/media-notes/WAPBrochure.php.
2. "Risks and Benefits of Estrogen Plus Progestin in Healthy Postmenopausal Women, Principal Results From the Women's Health Initiative Randomized Controlled Trial." *JAMA.* 2002;288(3):321–333 "Writing Group for the Women's Health Initiative Investigators," http://jama.jamanetwork.com/article.aspx?articleid=195120.
3. Xianglan Zhang, MD, MPH; Xiao-Ou Shu, MD, PhD; Honglan Li, MD; Gong Yang, MD, MPH; Qi Li, MS, MD; Yu-Tang Gao, MD; Wei Zheng, MD, PhD. "Prospective Cohort Study of Soyfood Consumption and Risk of Bone Fracture Among Postmenopausal Women." *Arch Intern Med.* 2005;165:1890–1895.
4. Pawlowski JW1, Martin BR, McCabe GP, McCabe L, Jackson GS, Peacock M, Barnes S, Weaver CM. "Impact of equol-producing capacity and soy-isoflavone profiles of supplements on bone calcium retention in postmenopausal women: a randomized crossover trial." *Am J Clin Nutr.* 2015 Sep;102(3):695–703.
5. Ma DF, Qin LQ, Wang PY, Katoh R. "Soy isoflavone intake increases bone mineral density in the spine of menopausal women: Meta-analysis of randomized controlled trials." *Clinical Nutrition.* February 2008, vol. 27, Issue 1, 57–64.
6. Chen et al. "Beneficial effect of soy isoflavones on bone mineral content was modified by years since menopause, body weight, and calcium intake: a double-blind, randomized, controlled trial." *Menopause.* 2004 May-Jun;11(3):246–54.
7. Massey LK. "Dietary animal and plant protein and human bone health: a whole foods approach." *J Nutr.* 2003 Mar;133(3):862S–865S.
8. Remer T, Manz F. "Estimation of the renal net acid excretion by adults consuming diets containing variable amounts of protein." *Am J Clin Nutr.* 1994;59:1356–1361.
9. Nutrient data for this listing was provided by USDA SR-21.
10. U.S. Department of Health and Human Services. *Bone Health and Osteoporosis: A Report of the Surgeon General.* Rockville, MD: U.S. Department of Health and Human Services, Office of the Surgeon General; 2004.
11. 144 Feskanich D, Willett WC, Stampfer MJ, Colditz GA. "Milk, dietary calcium, and bone fractures in women: a 12-year prospective study." *Am J Publ Health.* 1997;87:992–997.

12. Benjamin J. Abelow, Theodore R. Holford and Karl L. Insogna. "Cross-cultural association between dietary animal protein and hip fracture: A hypothesis." *Calcif Tissue Int.* 1992 Jan;50(1):14–8.

Chapter Twelve: Rumor: Asians Don't Eat Much Soy

1. *The Whole Soy Story: The Dark Side of America's Favorite Health Food.* New Trends Publishing, Kaayla T. Daniel 2005.
2. Messina M, Nagata C, Wu AH. "Estimated Asian adult soy protein and isoflavone intakes." *Nutr Cancer.* 2006;55(1):1–12.
3. The Okinawa Centenarian Study (OCS), okicent.org/study.html.
4. C. Nagata et al., "Association of Diet with the Onset of Menopause in Japanese Women." *Am J Epidemiol* 152, no. 9 (1 Nov 2000): 863–867.
5. "Response To Misleading Article About Soy In *Mothering Magazine.*" *John Robbins*,-foodrevolution.org/mothering.htm.

Chapter Thirteen: Soy Bashers: Who Are They Really Hurting?

1. Gregor Johann Mendel (1822–1884) Austrian scientist and Augustinian friar, considered to be the founder of the new science of genetics.
2. "Livestock's long shadow, Environmental issues and options," H. Steinfeld, P. Gerber, T. Wassenaar, V. Castel, M. Rosales, C. de Haan, 2006, 390.
3. Mora, C (2013). "The projected timing of climate departure from recent variability." *Nature* 502: 183–187.
4. IPCC, "Summary for Policymakers", Detection and Attribution of Climate Change, «It is extremely likely that human influence has been the dominant cause of the observed warming since the mid-20th century» (page 15) and «In this Summary for Policymakers, the following terms have been used to indicate the assessed likelihood of an outcome or a result: (...) extremely likely: 95–100%» (page 2)., in IPCC AR5 WG1 2013.
5. Eshel, Gidon, Pamela A. Martin, 2006: Diet, Energy, and Global Warming. *Earth Interact.*, 10, 1–17.
6. Leitzmann C. "Nutrition ecology: the contribution of vegetarian diets." Am J Clin Nutr. 2003 Sep;78(3 Suppl):657S-659S.
7. H. Steinfeld, P. Gerber, T. Wassenaar, V. Castel, M. Rosales, C. de Haan. "Livestock's long shadow, Environmental issues and options," 2006.
8. Animal products are the only source of dietary cholesterol.
9. Tropical oils, such as palm and coconut oil are the only plant foods that contain significant amounts of saturated fat.

Chapter Fourteen: Scientific Research Defined and Explained

1. Meredith Cohn, "Alternatives to Animal Testing Gaining Ground" http://articles.baltimoresun.com/2010-08-26/health/bs-hs-animal-testing-20100826_1_animal-testing-animal-welfare-act-researchers *The Baltimore Sun* August 2010.
2. Carbone, Larry. *What Animals Want.* Oxford University Press, 2004, p. 26.

3. PCRM Position Paper on Animal Research, July 21, 2010, pcrm.org/resch/anexp/position.html.
4. Chalmers TC, Smith H Jr, Blackburn B, Silverman B, Schroeder B, Reitman D, Ambroz A (1981). "A method for assessing the quality of a randomized control trial." *Control Clin Trials* 2 (1): 31–49.
5. Moher D, Hopewell S, Schulz KF, Montori V, Gøtzsche PC, Devereaux PJ, Elbourne D, Egger M, Altman DG (2010). "CONSORT 2010 explanation and elaboration: updated guidelines for reporting parallel group randomised trials." *Br Med J.* 340: c869.

Chapter Fifteen: So What *Is* the Skinny on Soy?

1. wnd.com/2006/12/39253/.
2. Gonda, T. *Estrogen Linked To Breast Cancer.* University Of Queensland (2007, August 24).
3. "Food Phytates" Rukma Reddy and Shridhar Sathe, CRC Press, 2001.
4. Layrisse M, García-Casal MN, Solano L, Baron M, Arguello F, Llovera D, Ramirez J, Leets I, Tropper E. "The role of vitamin A on the inhibition of non-heme iron absorption." *Nutr Biochem* 1997; 8: 61–7.
5. (7 U.S.C. 6401-6417) (Subtitle H of Title XIX of P.L. 101-624, called the Fluid Act)
6. *The Food Revolution*, John Robbins, Conari Press 2001.
7. "The Pew Initiative on Food and Biotechnology," February 2007, http://pewagbiotech.org/resources/issuebriefs/geneflow.pdf.

The Soyfoods Pantry

1. Bhagwat, S., Haytowitz, DB, and Holden, JM. 2008. USDA Database for the Isoflavone Content of Selected Foods, Release 2.0. U.S. Department of Agriculture, Agricultural Research Service, Nutrient Data Laboratory Home Page: ars.usda.gov/nutrientdata/isoflav.